INQUIRY AS INQUIRY:
A LOGIC OF SCIENTIFIC DISCOVERY

JAAKKO HINTIKKA SELECTED PAPERS

VOLUME 5

1. *Ludwig Wittgenstein*. Half-Truths and One-and-a-Half-Truths. 1996 ISBN 0-7923-4091-4
2. *Lingua Universalis vs. Calculus Ratiocinator.* An Ultimate Presupposition of Twentieth-Century Philosophy. 1996 ISBN 0-7923-4246-1
3. *Language, Truth and Logic in Mathematics.* 1997 ISBN 0-7923-4766-8
4. *Paradigms for Language Theory and Other Essays.* 1997 ISBN 0-7923-4780-3
5. *Inquiry as Inquiry. A Logic of Scientific Discovery.* 1999 ISBN 0-7923-5477-X

JAAKKO HINTIKKA
Boston University

INQUIRY AS INQUIRY: A LOGIC OF SCIENTIFIC DISCOVERY

KLUWER ACADEMIC PUBLISHERS
DORDRECHT / BOSTON / LONDON

A C.I.P. Catalogue record for this book is available from the Library of Congress.

ISBN 0-7923-5477-X

Published by Kluwer Academic Publishers,
P.O. Box 17, 3300 AA Dordrecht, The Netherlands.

Sold and distributed in North, Central and South America
by Kluwer Academic Publishers,
101 Philip Drive, Norwell, MA 02061, U.S.A.

In all other countries, sold and distributed
by Kluwer Academic Publishers,
P.O. Box 322, 3300 AH Dordrecht, The Netherlands.

Printed on acid-free paper

Printed in the Netherlands.

TABLE OF CONTENTS

ORIGIN OF THE ESSAYS vii

INTRODUCTION ix

1. "Is Logic the Key to all Good Reasoning?" 1

2. "The Role of Logic in Argumentation" 25

3. (with Ilpo Halonen and Arto Mutanen) "Interrogative Logic as a General Theory of Reasoning" 47

4. "What Is Abduction? The Fundamental Problem of Contemporary Epistemology" 91

5. "True and False Logics of Scientific Discovery" 115

6. "A Spectrum of Logic of Questioning" 127

7. "What Is the Logic of Experimental Inquiry?" 143

8. "The Concept of Induction in the Light of the Interrogative Approach to Inquiry" 161

9. (with Ilpo Halonen) "Semantics and Pragmatics for Why-Questions" 183

10. "The Varieties of Information and Scientific Explanation" 205

11. "On the Incommensurability of Theories" 227

12. "Theory-Ladenness of Observations as a Test Case of Kuhn's Approach to Scientific Inquiry" 241

13. "Ramsey Sentences and the Meaning of Quantifiers" 251

14. "Towards a General Theory of Identifiability" 267

ORIGIN OF THE ESSAYS

The following list indicates the first publication forums of the different essays included in the present volume (the forthcoming publication forum, if an essay appears here for the first time):

1. "Is Logic the Key to all Good Reasoning?," forthcoming.
2. "The Role of Logic in Argumentation," *The Monist* vol. 72, no. 1 (1989), pp. 3-24.
3. (with Ilpo Halonen and Arto Mutanen) "Interrogative Logic as a General Theory of Reasoning," forthcoming in R. Johnson and J. Woods, editors, *Handbook of Applied Logic*, Kluwer Academic, Dordrecht.
4. "What Is Abduction? The Fundamental Problem of Contemporary Epistemology," *Transactions of the Charles S. Peirce Society* vol. 34, (1998), pp. 503-533.
5. "True and False Logics of Scientific Discovery," *Communication and Cognition* vol. 18 (1985), pp. 3-14.
6. "A Spectrum of Logic of Questioning," *Philosophica* vol. 35 (1985), pp. 135-150.
7. "What Is the Logic of Experimental Inquiry?," *Synthese* vol. 74 (1988), pp. 173-190.
8. "The Concept of Induction in the Light of the Interrogative Approach to Inquiry," in *Inference, Explanation and Other Frustrations: Essays in the Philosophy of Science*, ed. by John Earman, U. of California Press, 1993, pp. 23-43.
9. (with Ilpo Halonen) "Semantics and Pragmatics for Why-Questions," *Journal of Philosophy* vol. 92 (1995), pp. 636-657.
10. "The Varieties of Information and Scientific Explanation," in B. van Rootselaar and J.F. Staal, editors, *Logic, Methodology and Philosophy of Science III,* North-Holland (Elsevier), Amsterdam, 1968, pp. 311-331.
11. "On the Incommensurability of Theories," *Philosophy of Science* vol. 55 (1988), pp. 25-38.
12. "Theory-Ladenness of Observations as a Test Case of Kuhn's Approach to Scientific Inquiry," in *PSA 1992*, ed. by David Hull et al., vol. 1, Philosophy of Science Association, East Lansing, MI, 1992, pp. 277-286.
13. "Ramsey sentences and the Meaning of Quantifiers," *Philosophy of Science* vol. 65, (June 1998), pp. 289-305.
14. "Towards a General Theory of Identifiability," in *Definitions and Definability*, ed. by James Fetzer et al., Kluwer Academic, Dordrecht, 1991, pp. 161-183.

All the previously published essays are reproduced here with the permission of the respective copyright owners, if any. These permissions are most gratefully acknowledged.
Thanks are also due to the Editors of the volumes in which these articles appeared previously.

INTRODUCTION

This volume has been singularly frustrating for me to edit. I am also painfully aware of the reason why. I am trying to do far too much in a single volume. I am trying to reform, or at least show how to reform, much of our epistemology and our general ("formal") philosophy of science. Such a task would require a longer sustained argument than can be provided by a single book, in at least two respects. On the one hand, the basic ideas of the new approach should be explained and motivated and their nature should be spelled out, including the conceptual tools needed to express them. On the other hand, the value of the new approach should be shown by applying them to a number of important representative problems.

There does not seem to be any *a priori* reason why such a task could be accomplished by a collection of studies like the present one. Admittedly, such a presentation imposes greater requirements on the reader than an integrated book. It is like presenting to the reader, not a ready-make picture, but rather a jigsaw puzzle to be completed by him or her.

Unfortunately, the picture I would like to paint is too large to be presented to the reader even in the form of a jigsaw puzzle of ideas. For that purpose, my basic ideas should have been explained more fully, more studies should have been carried out than has been possible, and those that I am presenting here should have been carried out to a greater depth and their common denominators explained much more patiently.

Hence what the reader will find here is an imperfect jigsaw puzzle from which many pieces are still missing. My explanation, if not an excuse, is that it would probably have taken another lifetime's work to present to you a completed picture of the kind I have in mind.

What I must try to do in this introduction is hence to help the reader by indicating what the hoped-for overall picture is intended to look like, how the pieces displayed here hang together, how they find their place in the total structure, and what some of the missing pieces are like. I am fully convinced that the missing tiles do exist for some cases because I have already found them but cannot present them in this volume.

Even though I referred to the approach advocated in the central essays of this volume as "new," its leading idea is not. This idea is as old as Socrates, and hence older than most of our familiar epistemology and logic. It is the idea of knowledge-seeking by questioning or, more accurately, of all rational knowledge-seeking as implicit or explicit questioning. I am using the phrase "inquiry as inquiry" to express the idea. For what my leading idea is is precisely an assimilation of all rational inquiry in the generic sense of searching for information or knowledge to inquiry in the etymological sense, that is, to a process of querying, or interrogation.

Needless to say, Socrates has not been the only philosopher to consider inquiry as inquiry. It would not be difficult to persuade a Collingwood or a Gadamer that each and every proposition figuring in rational inquiry can be thought of as an answer to a question. Unfortunately, neither one of these two philosophers provides us with any real argument for this position or anything like an explicit "logic of questions and answers," in spite of their prominent use of this very phrase. In the essay "What Is Abduction?" I put forward an argument to the effect that if a reasoner is rational, that is to say, if his or her argumentative steps can be rationally evaluated, then they might as well be considered as answers to his or her question. For the presuppositions of rational evaluation are the same ones that make it possible to think of any new item information as being introduced as an answer to a question. The main exception to the need of doing so is the one first noted by Aristotle, viz. an answer that is (as we are wont to say) logically implied by earlier answers (together with one's initial premises).

The logic of epistemology is therefore the logic of questions and answers — and of question-answer sequences. But the earlier forms of the theory of question-answer sequences are not adequate to the task. This prompted the paper "Interrogative Logic as the General Theory of Reasoning" (written jointly with Ilpo Halonen and Arto Mutanen) where an outline of an explicit theory of question-answer logic is presented. Alas, even within this survey paper many jigsaw puzzle pieces are still missing. Among them are most of the ingredients of a state-of-the-art epistemic logic. For it quickly turns out that while several of the basic features of a theory of questions and answers can be discussed by means of ordinary first-order logic, eventually the epistemic element in questions and answers must be acknowledged. To study this component adequately the development of an adequate epistemic logic is needed. This development is one of the many component tasks in the intended total reform of epistemology that cannot unfortunately be represented in this volume. Only a brief sketch of such a logic is presented here.

But what is the payoff of such an approach? Is it merely an old theory in new bottles? Perhaps the most important new insight that is yielded by the interrogative approach concerns the role of deductive logic in the context of an overall empirical inquiry. Needless to say, asking a question and receiving an answer (that is, an interrogative move) is radically different from a step logical deduction (logical inference move). However, from a strategic point of view the two steps are parallel, in the sense that the principles guiding the choice of the best questions to ask are (partly but importantly) analogous with the strategic principles guiding the choice of the best logical inferences one can draw from given premises.

Perhaps this result is not so new an idea, either. What it shows is that strategically speaking deductive logic is in the last analysis the guide to all good reasoning. I have called such a conception the Sherlock Holmes conception of logic. It can be taken to vindicate the great detective's claims to have solved his cases by "deductions," "inferences," or "logic." This idea is explained in easily accessible terms *inter alia* in the first two essays of this volume. The essay "The Role of Logic in Argumentation" is included partly for historical reasons, (being the first presentation of the analogy) even though it deals only with a part of the strategic analogy between questioning and deduction.

Another main insight that the interrogative approach yields is the possibility of a rational and even logical theory of discovery. Or, rather, I should say "actuality" instead of "possibility." For the logic of question-answer sequences already contains as one of its components a logical theory of discovery. Indeed, a theory of pure discovery is obtained as an extreme case by assuming that all of the answers that an inquirer receives are true and known to be true. Then the problem of justification becomes redundant, and yet we can develop an interesting, rich theory of discovery whose centerpiece is the problem of optimal question selection. There may not be any mechanical rules for the optimal question selection but that does not make the problem of discovery any less rational or even logical.

Philosophers used to maintain that only a "context of justification" can be dealt with by rational (discursive and logical) terms, whereas a "context of discovery" must be left to the mercies of intuition and serendipity. The true situation turns out to be an almost diametrically opposite one. A logic of pure discovery — that is, questioning with invariably true answers — has a beautifully clear structure. In contrast, a logic of justification will involve the complicated problem of evaluating different answers by different answerers, and will therefore end up being much messier than a logic of discovery.

Once again, an epistemological discovery prompted by the interrogative approach is better in step with common sense than recent philosophical preconceptions. For undoubtedly you would for instance rather base your argument in a court of law on the answers of an unimpeachable witness than those of a witness whose responses have to be tested against collateral information.

The "logic of discovery" that results from the interrogative approach depends essentially on the class of answers that the inquirer is in the position to receive. Most philosophers have apparently assumed that for a scientific inquirer all the rock-bottom answers must be thought of as particular propositions. This assumption has led to the inductivist and to the hypothetico-deductive models of science. In reality, it is nevertheless totally unrealistic, as is illustrated among other things by the possibility of putting questions to nature in the form of experiments. An answer to an experimental question is typically a functional dependence between two variables, which can only be expressed in terms of dependent quantifiers, and hence not a particular proposition.

These matters are discussed in essays 5-6 below. Again, they are unfortunately only the first sketchy words on the subject, not the last ones.

Thus the logic of the experimental method can be spelled out and studied by reference to the interrogative approach. A beginning of such a study is made in the essay "What Is the Logic of Experimental Inquiry?" It has already led to two further insights of major importance. First, most actual dependence relationships established experimentally are only partial ones, holding only for a restricted range of the values of the controlled variable. The problem of extending and integrating such partial generalizations is a frequent and important kind of scientific problem whose general nature and whose history is only now beginning to be visible. It is discussed in my essay "The Concept of Induction in the Light of the Interrogative Approach to Inquiry."

The other interesting insight is that the conclusiveness conditions of answers to questions (not only experimental questions) are not entirely empirical in nature, but contain a conceptual element. In the case of experimental questions this conceptual element amounts to knowing what the function is mathematically speaking that an experiment has yielded as its initial output. I have discussed this matter in my paper "Knowledge of Functions in the Growth of Mathematical Knowledge" forthcoming in a volume entitled *The Growth of Mathematical Knowledge* edited by Herbert Breger and Emily Grosholz (Kluwer Academic, Dordrecht).

Another promising line of thought opened by the question-theoretical approach is to develop an interrogative analogue to the usual theory of definability. In the new theory it is not any longer asked whether (and how) certain concepts are fixed uniquely by their role in some given theory alone. It is now asked whether (and how) they are specified by the theory plus the answers the inquirer can receive. It turns out that this interrogative analogue to definability has been acknowledged, given a name, and studied by the practicing methodologists called econometricians and systems theorists. It is a generalization of what they call identifiability. Some of the basic features of a general logical theory of identifiability are discussed in the essay "Toward a General Theory of Identificability."

A part of the interest of the concept of identifiability is due to the fact that it shows that a concept can be empirically determined by other concepts and yet not reducible to them. In this respect it is similar to, but much more sharply defined, than the confused notion of supervenience that is undeservedly popular in these days.

A major application of the ideas which have motivated the interrogative approach concerns the theory of explanation. I have studied this subject in recent years in cooperation with Ilpo Halonen. Our results are too extensive to be included in this volume. What we have done is to indicate the question-theoretical basis of this study in the paper "Semantics and Pragmatics for Why-questions."

Even though the older paper "The Varieties of Information and Scientific Explanation" predates my interrogative approach to inquiry, it complements this approach by illuminating the role of the concept of information in the scientific process.

Information is the most important "epistemic utility" which helps to determine the payoffs and hence the strategies of interrogative games.

One cannot discuss general problems of the philosophy of science in these days without commenting on the claims of the so-called "New Philosophers of Science," such as Kuhn, Hanson, and Lakatos. In the course of years, I have become increasingly disenchanted with their New Philosophy of Science, even apart from the simple-minded uses of the ideas of Kuhn and Hanson by their self-appointed followers. The main concept Kuhn and his ilk are wielding are far to unsharp, not to say confused, to serve as means of reaching serious insights into scientific discovery and scientific process in general. In the essays on the incommensurability of theories and on the theory-ladenness of observations, I show by means of examples that the New Philosophers' claims concerning these two typical concepts are at best confused and superficial.

This will make (I hope) the outline of the puzzle picture I am dealing with clearer to the reader. I trust that it will also show the reader how much needs still to be filled in within those outlines. But even so, the picture that is beginning to emerge shows several truly important things about the epistemology of science and of the science of epistemology.

1

IS LOGIC THE KEY TO ALL GOOD REASONING?

The syllabi, the textbooks and the curriculum design in elementary logic, reasoning, argumentation and critical thinking are all currently ill-conceived and ill-executed. In a certain historical perspective, which is only slightly oversimplified, the same fate has befallen logic as an academic discipline has befallen ethics. Ethics began in the ancient Greece and more generally in the classical antiquity as a study of different forms of excellence. These types of excellence were not only moral in our sense, but included sundry forms of social, military and intellectual superiority. As Adkins (1960, pp. 32-33) sums up the Homeric sense of the word, to be a good or virtuous man, that is,

> to be *agathos* one must be brave, skillful, and successful in war and in peace; and one must possess the wealth and ... the leisure which are at once the necessary conditions for the development of these skills and the natural reward of their successful employment.

Even though later in the antiquity somewhat more emphasis came to be put on cooperative virtues, the force of the key words *agathos, arete, virtus,* etc. was to indicate social excellence. A virtuous man was literally a virtuoso performer on the stage of communal life.

Alas, a funny — or, rather, sad — thing happened on the way to Victorian morality. Ethicists' emphasis switched from social and moral excellence to the avoidance of moral mistakes, to the preservation of one's virtue. As Cornford (1945, p. vi) puts it, a naïve modern reader

> who opened Jowett's version [of Plato's *Republic*] at random and lighted on the statement (at 549b) that the best guardian for a man's "virtue" is "philosophy tempered with music," might run away with the idea that in order to avoid irregular relations with women, he had better play the violin at the intervals of studying metaphysics.

Cornford is parodying the mistake of assimilating *arete, logos* and *musike* to what our words *virtue, philosophy* and *music* mean.

While no real scholar of the history of ethics is likely to make such a mistake, it seems to me that similar mistakes are being committed in our contemporary teaching of logic and reasoning. For it is unmistakable that a similar defensive attitude has taken over people's (or at least philosophers') conceptions of logic and reasoning. Logic began in the hands of Aristotle as an art of interrogative reasoning, modeled on the Socrate *elenchus*, so much so that he is still in his two *Analytics* assuming a dialectical or at least interrogative framework for all of the kinds of reasoning that he is considering there. And Aristotle's original theory, presented in the *Topics*, is firmly aimed at identifying and cultivating excellence in the interrogative games that were thought of by him as the medium of all reasoning.

But on the way to Boole and Frege, the idea of logic as the study of excellence in reasoning was gradually forgotten. It was largely replaced by an emphasis on the infallibility of logical reasoning, on the avoidance of mistakes in logic, not to say on the preservation of one's logical virtue. Frege's creation of symbolic logic is an admirable intellectual achievement, but no one is likely to claim that his formalization, including his notation, is conducive of brilliant reasoning even in pure deductive logic.

This insecure attitude is amply in evidence in our contemporary textbooks of informal logic, reasoning and argumentation. In each such textbook there typically are chapters on what are known as fallacies. These chapters are likely to be the sorriest anachronisms you ever saw. Many of the so-called fallacies are not mistaken inferences at all, but mistakes in questioning (fallacy of many questions, begging the question), mistakes of confusing one kind of inquiry with another one (as in the so-called *ad hominem* fallacy), or simply instances of bad strategies in reasoning. The defensiveness of such textbooks is illustrated by the title of a recent primer called *Logic: The Art of Verbal Self-defense*. How insecure can you get?

But we have here something much more serious than a wrong overall attitude. The metatheory of logic has been developed in a way that is not focused on excellence in reasoning. In order to gain an overview on the situation, it is useful to make a distinction between *definitory* rules and *strategic* rules of any goal-directed activity that can be conceptualized as a game. For instance, the definitory rules of chess tell you how chessmen may be moved on the board, what counts as checking and checkmating, etc. These rules define the game of chess. If someone tries to move a piece in violation of the definitory rules, it is not a genuine move. It has to be taken back, it is null and void. In contrast, the strategic rules (or principles) of chess tell you how to make the moves, in the sense of telling which of the numerous admissible moves in a given situation it is advisable to make. They tell you what is good play and what is bad play, if not absolutely, then at least relatively.

But if you only know the definitory rules of chess, you cannot say that you are a chessplayer. No one would deign to play a game of chess against you. You cannot even say that you know how to play chess. For the purpose, you must have some grasp of the strategic principles of chess.

This distinction can obviously be generalized to all goal-directed activities which can be conceptualized as games in the sense of the mathematical theory of games. (Game theory was founded by von Neumann and Morgenstern in 1944. For a state-of-the-art survey, see Eatwell et al., editors, 1989.) In any such activity, we can specify the moves that are admissible in different possible situations in abstraction to what a game theorist would call the payoffs in some play of the game. The payoffs determine the rewards and punishments that accrue to each player when the game is played to the bitter or sweet end. These payoffs clearly determine the value of different strategies that the players may use, but admissible moves can — and must — be specified independently of the payoffs.

The details of the game-theoretical treatment of sundry goal-directed activities are not important. However some of the main concepts and ideas are relevant to my line of thought. The crucial concept in game theory is the concept of strategy, which there means something that in colloquial terms would be called a *complete* strategy. It is a rule that tells a player what to do in any conceivable situation that might come up in the course of a game. Then the entire game can be reduced to the choice of a strategy by each player. These choices determine completely the course of the play and hence determine the payoffs. And these payoffs specify the value of the strategies chosen. Strategic rules hence concern in principle the choice of such complete strategies.

How do these ideas apply to the game of deductive logic? One thing is crystal clear. *The so-called rules of inference are definitory rules, not strategic ones*. At each stage of a deductive argument, there are normally several propositions that can be used as premises of valid deductive inferences. The so-called rules of inference will tell you which of these alternative applications of the rules of inference are admissible. They do not say anything as to which of these rule applications one ought to make or which ones are better than others. For that purpose you need rules of an entirely different kind, viz. strategic rules. The so-called rules of inference are merely permissive. They are rules for avoiding fallacies. They are not "laws of thought" either in the sense that they would tell us how people actually draw inferences or in the sense that they would tell us how we ought to draw inferences.

Yet in introductory logic courses a lion's share of attention is devoted to the *definitory* rules of inference. Students are told what these "rules of inference" are and they are taught to apply them correctly. Needless to say, part of this enterprise is unavoidable in logic teaching. Without knowing the definitory rules of a game, one cannot understand its strategic rules, either. But if all or most of the attention is devoted to the definitory rules, the enterprise will remain predominantly defensive, an exercise in avoiding fallacies rather than discovering proofs or finding out new truths by means of deductive inferences.

This is analogous to the fallacy of conceiving of ethics as the art of avoiding moral mistakes. In concentrating their teaching on the so-called rules of inference logic instructors are merely training their students in how to maintain their logical virtue, not how to reason well. Textbook discussions of different kinds of fallacies sometimes remind me of the elaborate and detailed taxonomies of different sins found in old handbooks of moral theology written for the benefit of parish priests who had to be ready for confessions of any sort of common or uncommon transgression.

This description of logic instructor's predicament might not seem alarming. Why don't the authors of the next generation of logic texts simply add to the definitory rules of the game logic, to wit, to the rules of inference, some suitable strategic rules? This is of course what ought to be done. It is more easily said than done, however. For one thing, the relevant strategic rules will have to be radically different from what most philosophers are likely to think first. Being strategic rules, they will have to refer in the first place to entire strategies, or at least to partial strategies. They cannot normally be formulated by reference to particular moves. They do not tell us what move to make in some particular situation, except insofar as that move is a part of some overall strategy. This makes of course good sense. If you ask me, "Will I find my way to South Station if I turn left in the next corner?", no simple answer "yes" or "no" is possible. The answer depends on what you will do in subsequent corners, that is, the rest of the strategy you are following. In books on chessgames, moves are often characterized as good, bad or chancy. But if I am playing against Kasparov and happen to make a move that would be labeled good, it would not have a predictive value as to who will eventually win, for my overall strategy will be inferior to Kasparov's.

Hence the strategic rules of logic must be of a radically different kind from the definitory rules, that is, from the so-called rules of inference. They will have to pertain to strategies, that is to say, sequences of moves and not to moves taken one by one, as the definitory rules do. Moreover, they must pertain to several alternative possible sequences of moves. There is a book on Kasparov's chess strategies (Padolfini, 1986) which discusses different games he has played over the years. It is full of analyses of possible positions which did not come up in the games Kasparov actually played but which could have arisen if his opponent had made a different move. The author argues perceptively that the greatness of Kasparov as a chess strategist cannot be spelled out without considering what he would have done in different unactualized situations. This is typical of strategic skills. The strategic rules of logic, too, cannot be formulated move-by-move, in contradistinction to definitory moves.

This point is worth emphasizing because there is an extremely strong tendency among the theorist of both deductive and non-deductive reasoning to try to formulate their rules as definitory ones, that is to construct sets of definitory rules of some new logic that captures certain strategic ideas. Such attempts are very dubious theoretically as well as practically. It seems to me that such developments as e.g. theories of belief change (cf. Gärdenfors, 1988), nonmonotonic logics, default logics, etc. should be subjected to a much sharper critical scrutiny from a strategic viewpoint than has been the case so far.

Furthermore, it can be shown that in logic the optimal strategic rules, even if one tried to formulate them move by move, cannot be recursive (mechanical). Hence they should perhaps be called principles rather than rules. In teaching, they might take the form of things that the student should learn to keep in mind in constructing her or his arguments.

From the fact that the so-called rules of inference in logic are merely permissive an interesting further conclusion follows. This observation shows that insofar as fallacies are thought of as violations of the definitory rules of logic and reasoning, there cannot be a separate study of fallacies apart from the study of logic and its rules of inference *simpliciter*. For what permissive rules do is precisely to define what is fallacious and what is not in inference and more generally rule-governed reasoning. There can be a separate, nonpsychological theory of fallacies only so far as fallacies are thought of as strategic mistakes, not violations of the "rules of inference" of logic. And it follows from what has been said that the theory of strategic fallacies cannot be built in the last analysis as a theory of fallacies in inference, more generally as a theory of how particular moves are made in logic or reasoning. For it was seen that in general utilities can in the first place be associated only with entire strategies, not with particular moves. This result has important implications for the theory of strategic fallacies — which is the only kind of independent theory of fallacies. The motto of every fallacy theorist should be Talleyrand's (probably apocryphal) comment when he was asked about the kidnapping and execution of duc d'Enghien by Napoleon's agents: "It is worse than a crime, it was a mistake." In other words, it was not only a violation of the definitory rules of international law; it was a violation of Napoleon's own optimal strategy in the game of *Realpolitik*.

As was mentioned, Aristotle was very much concerned with the strategic aspects of reasoning, which he conceived of as an interrogative (dialectical) process. Ironically, the first definitory rules of logical inference resulted from Aristotle's very strategic interest in strategies of questioning. (Cf. here Hintikka, 1996.) In all such questioning, one of the most important strategic considerations is to anticipate the answers that the inquirer is likely to receive to her or his questions. (Just think of cross-examination in a court of law. Lawyers' folklore even includes the old saw that one should never ask a question in cross-examination if one does not know what the answer will be.) This interest in the predictability of answers led Aristotle to pay special attention to a class of answers that are completely predictable. They are the ones that are, as we would put it, logically implied by the interlocutor's earlier answers. Aristotle noticed the peculiar status of such answers and began to study them systematically, thus starting the science of logic.

But logic remained a part of an interrogative *elenchus*-like process. Right in the middle of explaining the nature of logically necessary inferences (that is to say, question-answer steps where the answer is implied by earlier answers) Aristotle suddenly catches himself and warns that such inferences are not the only steps of interrogative inquiry.

> Yet one might perhaps wonder what purpose there could be in asking about such items if it is not necessary for the conclusion to be the case — you might as well ask an

> arbitrary question and then state the conclusion. But we should ask questions not on the grounds that the conclusion will be necessary because the [earlier] answers necessitate the conclusion, but rather because it is necessary for the person who accepts the conclusion [i.e. answers the question in the proposed way] to state it — and to state it truly if they hold truly. (*An. Post.* A 6, 75a22-27; I have modified Barnes' translation.)

Most recent translators and commentators have had no inkling of what is going on in this passage. Out of desperation, some even try to insert a completely unsupported reference to dialectical reasoning. Yet as soon as we realize that Aristotle is treating the entire inquiry, including deductive steps, as an interrogative process, what he says becomes crystal clear. What this passage shows vividly is that logical inferences were for Aristotle a subclass of answers to questions. The conclusion of a logical inference was still an answer to a question. It also shows that what distinguishes them from any old answer to the inquirer's question is that the conclusion is necessitated by the earlier answers.

So far I have dealt almost exclusively with deductive reasoning. However, there is a way of extending practically everything I have said to rational inquiry in general, that is, to generalize our deductive logic so as to become the theory of all reasoning, ampliative as well as deductive. For the purpose, we only need to go back to the source of logic and borrow a page from Aristotle. All we need to do is to emulate Aristotle (see above) and to consider all reasoning as an interrogative process. Logical inferences will then constitute a subclass of the steps in such an inquiry, the other steps being interrogative (i.e. question-answer) moves. Historically speaking, of course what Aristotle originally did was not to distinguish conclusions of logical inferences from other answers to questions. As we saw, he only later came to pay special attention to such questions whose answers are predetermined and where we can for all theoretical — or at least nonstrategic — purposes disregard the question and consider only the answer.

The motivation of such an interrogative approach to reasoning in general might be explained as follows:

Consider a line of thought that is represented by a sequence of statements. What is required in order for us to be able to say that it is a line of *reasoning*, and not just a random sequence of thoughts? The obvious requirement is that it must be capable of being rationally evaluated as representing better or worse reasoning. But what is required for such an evaluation? Consider some one step S in the line of thought. Now the information S codifies is either contained in the information codified by the earlier steps or else it is at least partially new. In the former case, the evaluation of the reasoner's performance turns on the rule she or he used (explicitly or implicitly) to reach S from those earlier steps.

In the latter case, in order to evaluate the reasoner's performance we must know where the new information came from. Let us call that source O. Moreover, the reasoner must know that and must have selected that particular source of information O in preference to others. In order to be able to evaluate the reasoner's performance, we must be able to know what other responses O might have yielded instead of S. (For the reasoner did not know ahead of time that the response would be S.) Also, if other sources of information were available to the reasoner, we must know what information they might have yielded if the reasoner had consulted them.

But, if we know all this, we might as well think of the new information as an answer by the reasoner addressed to O in preference of addressing questions to alternative sources of answers. Hence all ampliative reasoning can be construed as a question-answer process, at least insofar as the answers can always be trusted by the reasoner. These question-answer processes can be conceptualized as interrogative inquiries in the sense explained and studied in my earlier papers. (For a survey, see Hintikka, Halonen and Mutanen, forthcoming.)

The main fact that emerges from this interrogative approach to epistemology, including scientific inquiry is a close analogy between the two kinds of steps in interrogative inquiry, the logical (deductive) inference moves and the interrogative (question-answer) moves. For the time being, it suffices to restrict our attention on the logic side to what is known as *tableau* methods or alternatively as tree methods. They were both introduced in 1955 by E.W. Beth and by Jaakko Hintikka, respectively. The original exposition of the *tableau* method is Beth (1955), and an extensive exposition of the tree method is found in Smullyan (1968). The analogy between deduction and interrogation is formally speaking due to the fact that, in the same way as in a logical inference move we need premises that have already been reached to draw the conclusion, in the same way we need an already established presupposition for the purpose of asking the question. For instance, a disjunction

(1) $(S_1 \vee S_2)$

can serve (when it occurs on the left side) as the presupposition of the propositional question

Is it the case that S_1 or S_2?

Such an interrogative step looks like this

(2)

—	—
—	—
—	—
$(S_1 \vee S_2)$	
S_i	

Where i = 1 or 2. But it can (and sometimes must) also serve as the input into the deductive move of *tableau* building in which we divide the argument into two, considering tin the one of them S_1 and in the other S_2 as being true, like this:

(3)
$$\begin{array}{c|c} — & — \\ — & — \\ — & — \\ (S_1 \vee S_2) & \\ S_1 \mid S_2 & \end{array}$$

Here (2) is like (3) except that one of the two *subtableaux* originating from the move (3) is omitted. Of course, in asking a question we do not usually know which answer the advisee will give. Hence, strategically speaking we have to consider both possible answers, just as in (3).

Likewise, an existential statement

(4) $(\exists x)\, S[x]$

can serve either as the presupposition of the wh-question

(5) Which individual (say, x) is such that S[x]?

or else as the premise of the existential instantiation move

(6) $\dfrac{(\exists x)\, S[x]}{S[\alpha]}$

Notice that if an answer to (5) is available, it will be of the same form as the conclusion in (6), except that instead of the "dummy name" (or the name of an "arbitrary individual") α we have the name of an actual individual for instance

(7) S[b]

These remarks pertain to simple wh-questions only. It has been shown, however, how they can be extended to arbitrarily complex wh-questions, (see Hintikka, Halonen and Mutanen, forthcoming).

Again, in general we do not know sight unseen what the individual b is that our respondent is offering by way of an answer. Hence our strategic thinking steps involving simple wh-questions will have to be thought of along the same lines as existential instantiations of deductive logic.

In virtue of this analogy between interrogative moves and logical inference moves in the "game" of inquiry most of what was said above of logical reasoning can be extended to reasoning in general, including empirical inquiry. In particular, we find both in the philosophical theorizing about reasoning and in the teaching of reasoning argumentation and critical thinking the same mistake as was diagnosed above in the case of deductive reasoning and argumentation. The mistake is an almost exclusive emphasis on definitory rules of different rational goal-directed activities, at the expense of their strategic rules. In epistemology and philosophy of science this has led to a proliferation of various new logics, such as nonmonotonic logics, default logics, logics of belief change, inductive logics, logics based on circumscription, etc. There is nothing wrong with such logics as such, but they do not have the philosophical or practical significance generally claimed for them. They seek to imitate or generalize the trivial aspects of deductive logics, viz. the definitory rules of logic. They do not take seriously the strategic aspects of the games of reasoning. This is witnessed by the fact that they operate by and large in terms of step-by-step rules just like the definitory rules of deductive logic. For instance, the rules we find in the theory of rational belief change are typically one-step rules as to change our belief system in view of one new item of information. (See e.g. Gärdenfors, 1988.) They have nothing to say of strategies of belief change by means of the acquisition of further new information, in spite of the fact much of the relevance of new information lies in its role in facilitating the acquisition of further information, for instance by providing presuppositions for further questions. It is a serious mistake to consider the impact of new information without taking into account the possibilities this information opens for further inquiry. And yet this is what is being done in most of the recent theorizing about belief change.

In epistemology, the same mistake is rampant. One of the key terms in recent discussions in the theory of knowledge is *warrant*. (See e.g. Plantinga, 1993.) Now what a warrant is is the backing of one particular step in reasoning, as it were the collateral in the particular epistemic transaction. In the light of what has been said, the study of warrants can perhaps be the alpha of the study of reasoning, but it cannot be its omega. And so far as the theory of the pursuit of excellence in reasoning is concerned, it is not even the beta of that enterprise. A scientist is typically less interested in questions like "To what extent does the evidence I already have support my hypothesis?", or even "What does this evidence tell me about the world?" than in such questions as "How can I find new evidence to support or to reject my hypothesis?" or "What experiments and observations should I make to discover what the world is like?".

The same mistake is committed almost always in the teaching of reasoning, argumentation and critical thinking. Insofar as the teaching of these subjects is not construed as blind training, it is usually thought of as taking the form of conveying certain rules to the students. Moreover, these rules pertain to particular steps of reasoning, not to strategies of reasoning. Even in the best textbooks of critical thinking, such as

Toulmin, Ricke and Janik (1979), the emphasis is on the warrants of the particular moves an arguer makes in her or his line of reasoning. It may be that one can teach safe reasoning in this way, but there is no hope of so teaching excellence in reasoning. More generally speaking, in current textbooks of reasoning and argumentation, arguments are evaluated almost exclusively in terms of their persuasiveness rather than in terms of the new knowledge they produce.

In view of such mistakes in philosophers' overall approach to reasoning, it is no wonder that even the results promised by typical textbooks of logic and reasoning are pitiful. The horror story should stare everyone in the face who opens an introductory text in reasoning, argumentation or logic for philosophers. It is therefore scarcely necessary for me to try to document this scandal *ad nauseam*. Let a couple of examples suffice. In one logic textbook after 388 pages the students are just about able to manage the following brilliant deduction:

> James and Jimmy are not the same individual. James is my brother, and so is Jimmy. Therefore, I have at least two brothers.

It is also instructive to see that some of the more perceptive textbook writers unmistakably have a bad conscience on this score. Their penance is to try to create the impression of dealing with surprising consequence relations. A case in point is the tired old example:

(8) Everybody loves a lover. Alma loves herself. Therefore the Baron loves Alma.

This is a bare-faced cheat, however. The appearance of surprise is created by a mistranslation, that is, by translating the verbal example (8) into the formal notation in an unintended way. For in ordinary discourse "a lover" means in effect "someone else's lover," and if this is taken into account, (8) turns out not to be a valid inference. The (mildly) interesting task here would be to point out that invalidity and diagnose its reasons.

Most of the other nontrivial-sounding examples in logic texts are similar cheap tricks. After more than forty years, one of the few introductory logic texts for philosophers which uses nontrivial application-oriented examples is Suppes (1957). (Cf. also Salmon, 1989.)

Not surprisingly, the material that in philosophers' logic texts takes up to 400 pages are in introductory undergraduate logic texts for mathematicians typically covered in some 40-50 pages.

Again, similar no-brainers are found in the textbooks for reasoning and critical thinking courses. A brand-new reasoning text culminates after 346 pages in the remarkable insight that "the willingness to keep the dialogue open is the hallmark of the rational person." Would a person who did not believe this take a course in reasoning and argumentation in the first place? In the same text, a definition of fallacy is used that does not apply to some of the best-known fallacies discussed by Aristotle.

Putting forward such criticisms is admittedly a cheap thrill. It may sound fine and dandy, but can I really suggest guidelines as to what to aim at in teaching logic, reasoning and argumentation? I will try to convince you that such guidelines are available. Let us first return to the teaching of logic. There one indeed can say a number of things about how the strategic aspects of deductive reasoning should be taught. In a certain sense, not only has the main emphasis recently been on definitory rules rather than strategic rules, it has been on the wrong sets of definitory rules. What I mean by this is the following: The main overall theoretical development in logical theory is undoubtedly the model-theoretical interpretation of first-order logical inferences codified in the *tableau methods* or *tree methods* described above. The basic idea is to consider an attempt to show that F logically implies G as an experimental attempt to construct a description of a scenario in which F is true but G is false. If all the possible ways of doing so turn out to lead to an overt contradiction, then G is indeed a logical consequence of F. By an overt contradiction, I mean a situation in which the same sentence is said to be true and false. By using an idea first proposed by Henkin (1949) we can interpret this construction as an attempted model construction instead of the construction of a mere description of a countermodel, using sets of formulas as their own models.

Either way, a set of rules of first-order logic can be read almost directly from this idea. For instance, the steps (3) and (6) above instantiate such rules.

We could call this perspective on proofs by the *tableau* method or by the tree method the model-theoretical interpretation of cut-free logical proofs. It is of great interest, even though this interest has not fully caught the attention of philosophical logicians. Its proof-theoretical interest is shown by the fact that arguments ensuing from a completed *tableau* are in effect simply cut-free Gentzen-type sequent calculus proofs written upside down. (What is meant by cut-free proofs will be explained later.) For Gentzen methods, see Gentzen (1969). The model-theoretical interpretation obviously has a great theoretical interest in relating proof theory and model theory to each other. What is especially interesting here, the model-theoretical interpretation of cut-free proofs throws light on the strategy selection in constructing such proofs. The crucial steps in a search of such proofs, that is, in a *tableau* construction, are the introductions of new individuals through existential instantiation on the left side (and through universal instantiation on the right side). For only such steps change the configuration of individuals considered in the countermodel construction and add to the complexity of this configuration. Which instantiation is strategically best, that is, which new individuals is advisable for the reasoner to introduce, depends on the outcome of such repeated introductions. The ability to anticipate the configurations resulting from the different possible instantiations is clearly the secret of a good strategist in games of cut-free proof construction just as the ability to anticipate the outcome of future moves in chess is the gist of the strategic skills of grandmasters.[1]

It is worth noting here that, even when we are dealing with purely deductive reasoning, the ability to anticipate the outcome of a sequence of countermodel-building moves is not independent of the subject matter. In some walks of reasoning, the entire construction process is easier to envisage than in others. For instance, it is easier to do in geometry than in algebra. Moreover, familiarity of the possible configurations in some subject matter area may help the reasoner in trying to think sufficiently far ahead amidst the moves of the "game" of countermodel construction. Hence we have found an explanation why in real-life reasoning, even when it is purely deductive, familiarity with the subject matter can be strategically helpful.

Thus the model-theoretical interpretation opens extremely interesting perspectives into the strategies of deductive reasoning. Alas, these perspectives have not been adequately exploited in logic textbooks and in logic teaching, in spite of the recent popularity of *tableau* and tree methods in logic instruction.

But there is something much worse to be said of the emphasis on *tableau*-type and tree-type methods in logic teaching. Such cut-free deductive methods are not the only ones. Others are reached by allowing rules that do not satisfy the requirement that characterizes *tableau*-type methods, the requirement of the *subformula property*. What it says is that each new formula introduced in the course of a proof be a subformula of an earlier one. It is this subformula property that is the source of the elegance and easy interpretability (including model-theoretical interpretability) of *tableau* methods. The basic form of rules that violate the subformula property is the rule that allows the introduction of arbitrary disjunctions of the form (S ∨ ~S) as additional premises. Such sentences are undoubtedly true and hence acceptable as premises, but their connection with the initial premises, interim conclusions, and the intended ultimate conclusion is in principle completely free, as far as the definitory rules of logic are concerned. The corresponding Gentzen-type rule is sometimes known as the *cut* rule. This explains my earlier references to *tableau* and tree methods as cut-free ones.

In many treatments of first-order logic, such non-cut-free methods are implemented by means of conditional reasoning (reasoning from premises) rather than tautology-introducing rules. The two techniques are equivalent, but the cut-type rules are more perspicuous.

The first main systematic result in contemporary proof theory is Gentzen's first *Hauptsatz* which says that the cut rule and similar rules can always be eliminated. What can be proved by their means can be proved without them. This might seem to vindicate the use of cut-free methods, in that they show that they can do everything that other valid deductive methods can do in first-order logic. But this equality is so to speak merely definitory, not strategic. Cut-free methods enable us to prove anything that can be proved but not equally efficiently. The non-cut-free methods can simplify proofs enormously and in this way make proofs easier to find and to carry out. Indeed, for all their philosophical and other theoretical advantages, *cut-free methods are* in a perfectly good sense the slowest, that is to say, *the least efficient methods of logical inference* in first-order logic. To use them exclusively or almost exclusively in introductory logic teaching is thus a pedagogical absurdity, a telling example of the disastrous consequences of neglecting the strategic viewpoint in logic teaching. Cut-free systems of rules of inference may be well designed to understand the nature of logical inference, even though they are not even being currently used very well for that purpose. They may also be the rules of inference whose justification is optimally clear. But insofar as the aim of logic teaching is to bring the students to reason efficiently, to help them to achieve excellence in deductive reasoning, cut-free methods are the worst possible ones, if used alone.

In practice, fortunately, the use of cut-free rules is in typical textbooks supplemented by other rules, most often by rules of conditional reasoning. But even when they are used, their significance and their strategic role are not brought out.

In terms of the model-theoretical interpretation of first-order logical inferences, one can see what happens when cut-type rules are used, hopefully to simplify proofs. In such a rule, one introduces (S ∨ ~S) as an additional premise. What that means is that instead of building an experimental countermodel by introducing new individuals one by one, we build the countermodel in two possible ways and introduce in one fell swoop into the construction the entire structure described alternatively by S or by ~S. And this structure is described, not by listing its ingredient individuals one by one, by specifying their respective relations, etc., but as it were only verbally, as denizens of the models of the statements S and ~S, respectively.

Since one can in this way introduce a multitude of new individuals at once, with a predictable (on the basis of the choice of S) structure to boot, it is not surprising that by using the cut rule one can carry out first-order proofs much more efficiently than without it. At the same time, the cut rule makes it both easier and more difficult to plan one's proof strategically. The crucial task is of course still to anticipate the outcome of a logician's successive countermodel construction steps, no matter whether they introduce new individuals (and otherwise complicate the partial model so far constructed) piecemeal as in cut-free rules, or wholesale, as in the cut-rule type rules. The wholesale rules offer the deductive reasoner a far greater range of opportunities as to what to do, in the sense of how to continue the building of a putative countermodel. At the same time, this variety makes it more difficult to choose between different options. Furthermore, the complexity of the newly introduced complexes of introduced individuals makes it more difficult to anticipate the outcome, especially as the added complexity is in the cut rule only described, not actually displayed. This makes the mastery and the teaching of the powerful strategies opened up by the cut rule more difficult. However, this difficulty is a poor excuse for not trying to teach them. For if your aim is to cultivate excellence in deductive reasoning, the strategies that are opened by these powerful rules should be the focal point, or perhaps rather than the ultimate goal, of the teaching and learning of deductive logic.

The fact that in cut-free rules new configurations of individuals are introduced into deductive reasoning discursively rather than by a direct construction has had wide-ranging effects on people's ideas about logic. It has encouraged the idea that deductive reasoning including its strategies must (or at least should) be dealt with by means of verbal rules, such as symbolic and formal rules, rather than pictorial (iconic) ones. This is a fallacy. In reality, figurative language, such as the use of diagrams and figures in geometry, can be as explicit and accurate as a purely formal language in which geometrical reasoning is formulated ("formalized"). Once again we are dealing with a result that has pedagogical implications. For instance, the connection between first-order logic and the domino problems studied by Hao Wang (1993, especially Appendix A) could be used to illustrate various strategic and philosophical aspects of first-order deductive reasoning. This is possible because constructing a logical proof is more like solving a jigsaw puzzle than executing an algorithm.

But what does this have to do with reasoning in general? Almost everything, for much of what has been said can be extended to all rational inquiry by means of the interrogative approach to inquiry explained above. One can represent interrogative inquiry in a *tableau* form, and apply most of what was said of deductive *tableau* construction to interrogative inquiry. There nevertheless are some differences. I will first consider the case of pure discovery, that is, a situation in which all the available answers are known to be true.

(1) Obviously the left hand side must accommodate not only the sentences that are postulated to be true in the course of the experimental model construction. It must also accommodate all the answers.

If not all answers are true, the left side must include only such answers as have been received and have not been rejected ("bracketed").

(2) In a fully explicit form of interrogative inquiry, we must make the epistemic ingredient of the process explicit. This necessitates the use of epistemic logic rather than first-order logic. However, I will not here discuss this complication. It is largely needed for the purpose of using interrogative inquiry for the purpose of answering questions of different kinds, rather than establishing specific conclusions, which is what I am focusing my attention here. (By "answering questions," what is meant is of course answering initial "big" or principal questions by means of successive "small" or operative questions.)

Otherwise, the parallelism between logical inference steps and interrogative steps described above continues to hold. It has to be extended in one respect, however.

Applications of the cut-type rule that introduces $(S \vee \sim S)$ anywhere on the left side will now correspond to yes-or-no questions of the form

(S or ~S)?

The reason why they may be asked at any time and for any S is the fact that their presupposition is vacuous. For all the other types of questions their presuppositions have to be established first (i.e. have to be present on the left side) before they may be asked.

Pretty much everything that was said of the strategies of deductive reasoning can be extended to all reasoning, conceived of as interrogative reasoning. In the same way as in deductive reasoning we have to introduce the right new individuals through a judicious choice of existential instantiations, in the same way we have to ask the right wh-questions in order to introduce suitable individuals via answers to them. In both cases, the hands-on strategy selection is governed by one's ability to anticipate the future course of one's reasoning-as-countermodel-construction.

In one respect, my generalization of the *tableau* method nevertheless opens a dramatic new perspective. The reader may have felt somewhat disappointed by what I have so far offered you by way of strategic advice for deductive reasoning. Basically, I have only told you what to try to anticipate, not how to make particular moves. Is there any help in sight? Absolutely speaking there is not but relatively speaking there is an extremely important result in the offing here.

What is the true purpose of logic? It surely is not merely an art for art's sake, a mere game in the pejorative rather than the von Neumann sense of the word. It is still for us an organon, a tool for reasoning other than purely deductive reasoning. And this certainly ought to be true with vengeance about the teaching of logic.

But what is the contribution of logic to reasoning and inquiry? Apparently we find here a division of labor between deduction and inquiry, codified in the distinction between logical inference moves and interrogative moves. The contribution of logic is to be the medium of the logical inference moves. But their role is a relatively modest one. In a logical inference move, the inquirer is merely reshuffling the information reached at the earlier stages of inquiry, integrating its different items and bringing them to bear on each other. This role is comparable with what happens in a Platonic dialogue when after an extended exchange of questions and answers Socrates says "Let's now add our admissions together." The main difference is that in reality such recombination of answers may be needed in the course of the dialogue itself, in order to have the presuppositions of questions available, and cannot be postponed to the end of the dialogue. Such a role of logic is not insignificant, but it is clearly only the less important part of the story or, rather, part of the inquiry. Hence the role of logic and logic teaching in the overall training of young and old people to think and to reason well seems to be rather modest.

This conclusion nevertheless involves a fatal fallacy. It is an instance of the fallacy I have repeatedly criticized in this paper. This fallacy is to concentrate only on questions concerning the definitory rules of inquiry and to neglect its strategic aspects. For even though a logical inference move and an interrogative move are different from a definitory point of view, the inquirer faces a closely similar and in a sense identical question in both of them. As was seen, in deduction the inquirer must choose an already established sentence as a premise of a logical inference move (or as one half of the disjunction $(S \vee \sim S)$ to be introduced). Likewise, in an interrogative move the inquirer must choose a previously established sentence as the presupposition of a question (or else choose an S in the yes-or-no question "(S or ~S)?").

Now the range of possible choices is the same in the two cases given some particular situation in which the inquirer finds herself or himself (or itself, if the inquirer is an automaton). The previously established propositions are the same for both choices (and so is the range of possible choices of S).

What we are interested in here is the strategically optimal choice of the interrogative move. Which previously established sentence (or available option in a yes-or-no question) should you choose? I cannot give you an absolute answer, but I can give you a conditional one. *You should choose the same sentence that you would choose in your optimal deductive strategy*. In other words, disregarding for a moment certain less important qualifications, *your best strategies in* interrogative reasoning, which is a model of *reasoning in general, are the same as your best deductive strategies*.

This constitutes an affirmative answer to my title question. It is an enormously important result. It shows that logic in the strictest possible sense of the word is not only one ingredient in reasoning in general, or rather that it is that only from the perspective of definitory rules of the "game" of inquiry. It is strategically speaking the alpha and omega of all good reasoning. We can also now see that the underestimation of the role of logic in reasoning in general is based on the same mistake that penetrates a great deal of contemporary thinking in general, to wit, on a failure to appreciate the strategic aspects of thinking and reasoning.

In view of this importance of the result just reached, we might dub it the first main theorem of interrogative inquiry or perhaps a little more modestly, the *strategy theorem*. For a somewhat more explicit formulation of this theorem, see Hintikka, Halonen and Mutanen (forthcoming). The main idea underlying this result was explained (under somewhat restrictive assumptions) already in Hintikka (1989).

The strategic role of logic in interrogative reasoning which has thus been brought out is easy to appreciate in terms of particular examples. For instance, if one of the already established sentences is

(9) $(\exists x)S[x]$

then the inquirer can use it either as the premise of an existential instantiation leading to a sentence of the form

(10) S[α]

where α is a "dummy name" or a name of an unknown "arbitrary individual," logicians' counterpart to the John Does and Richard Roes of legal parlance, not the same of a *bona fide* individual. But (9) can also be used as the presupposition of the wh-question

(11) Who (say, x) is such that S[x]?

If an answer is available, it will be of the form

(12) S[b]

where b is a name of an actual individual. Thus (12) differs from (10) only by having an actual name instead of a dummy name as the substitution-value of the bound variable x. This can make quite a difference in the rest of the inquiry, for one can ask questions about particular "arbitrary individuals," as one can ask questions about real ones. But this difference is as it were definitory. If we think of the situation strategically, the inquirer does not know who or what the individual b is that will be specified as (a part of) the answer to the question (11). Hence the inquirer must be prepared to countenance any answer. Therefore the inquirer might as well consider, strategically speaking, an arbitrary individual as an answer to the question.

From this is follows that if some particular choice of (9) as the input into the next step in the inquirer's deductive reasoning is optimal in the sense of helping to close the *tableau* in the simplest possible way, the same choice of (9) as the presupposition of the next question to be asked is also likely to be optimal. In this sense, the strategic logic of inquiry is nearly the same as the strategic logic of deduction.

Similar remarks can be made about propositional questions vis-à-vis *tableau* rules for disjunctions and about presuppositionless yes-or-no questions vis-à-vis the cut rule.

What has been said requires a few nonessential qualifications.

(i) The parallelism between deduction and interrogation does not help us with all the strategic problems in inquiry. In particular, it does not help us in deciding whether to ask a question or to use its presupposition as the premise of a deductive inference. Other things being equal, it is at least as good and possibly better to ask a question than to perform the corresponding deductive tableau-building step. But sometimes an answer is not available to the question, and sometimes the inquirer does not know whether an answer is forthcoming.

(ii) In practice, the most important qualification is due to the assumption I tacitly made above, viz. that the reasoner is ignorant as to what the outcome of a question will be. Now a question is pointless if it is already known for sure what the answer will be. But an inquirer may not be sure and yet can make an educated guess at the answer. Such guesses need not be merely matters of luck. They may be backed up by probabilities of some sort or other. If so, the inquirer may improve her or his strategies from what the mere parallelism recommends by taking these probabilities into account.

Such probabilities can be incorporated into the deductive process by assigning suitable weights (probabilities) to different sentences of our language. However, those probabilities cannot be fully determined by *a priori* considerations.

(iii) The parallelism holds directly only for inquiries where the answers can always be treated as being true. If we try to cope with the kind of reasoning where answers have to be evaluated for their reliability, the parallelism breaks down. It remains to be investigated to what extent it can be restored by constructing a logical treatment of the types of inquiry where doubtful answers can be at least tentatively rejected ("bracketed"). Perhaps some suitable kind of default logic could help us here. Pending a solution to this problem, we can say only that the strategic role of logic in inquiry is clearest in the kind of reasoning that aims at the discovery of new truths rather than at the confirmation of old ones. Deductive logic can offer you strategic advice for discovery of new truths, not for evaluating *prima facie* truths already obtained.

(iv) A pedagogical qualification to the parallelism lies in the very strategic nature of the parallelism between deduction and interrogation. Since we are dealing with strategies, we are at bottom dealing with longer series of questions and/or logical inferences. From this it follows that the strategic parallelism I have been discussing will show up only in relatively complex situations. Hence it is not immediately obvious how it can be best utilized in concrete logic teaching, where the situations concretely considered must be relatively simple.

In spite of these not unimportant qualifications, the near-identity between the principles governing deductive strategies and those governing strategies of reasoning in general remains an important insight. It vindicates in effect the popular view of logic as the secret of all good reasoning which I have called the "Sherlock Holmes concept logic." In spite of this jocular locution, what we are dealing with here is highly interesting and important. For one thing, my findings belie once and for all the disdain in which logic is currently being held as a vehicle of discovery by the majority of philosophers.

Among other things, the results reached offer guidelines for logic teaching. What happens usually in introductory logic courses in philosophy departments is that students are trained to construct proofs by using some standard set of rules of inference. Such training is not useful for any conceivable purpose. It is of course important for the sake of logic that the student understand what is going on in first-order proofs, but that is an altogether different enterprise from learning how to find such proofs. For the practical purposes of arguing well deductively, sundry shortcut methods are much more useful than a strict adherence of any one set of rules of inference. For the purpose of understanding the nature of first-order reasoning, the model-theoretical interpretation of cut-free methods is crucial, of course combined with an understanding of the nature of the cut rule.

For the purpose of using the strategic resources of logic for the purpose of reasoning well outside deductive logic itself, the focus of attention in logic teaching should be in the principles of selecting the right application of the different possible rules to perform first in any given situation. Yet this topic is not even mentioned in most textbooks of logic. Furthermore, as a prerequisite of the task of anticipating the outcome of repeated rule applications, the model-theoretic ("iconic") aspects of deductive reasoning should be emphasized much more.

In sum, logic is much more than an inference machine. It is *ipso facto* a strategy theory for all reasoning insofar as it is aimed at discovery rather than confirmation or disconfirmation. Philosophers of science have quarreled about the question of whether a logic of scientific discovery is possible. The controversy can be put to rest. A logic of discovery is possible because it is actual. Strategically speaking, our ordinary logic can be viewed as such a logic. What remains to be done is to study logic from this viewpoint and to harness the results to the service of the teaching not only of logic but of reasoning in general.

NOTE

[1] The importance of introducing the right "auxiliary individuals" into one's thinking is familiar to all puzzle fans. A great many of them (especially so-called lateral thinking puzzles) depend for their solution on the solver's lighting upon the right new object, whether by inference, conjecture or by way of an answer to a question does not matter here.

Here is a familiar example:

> A woman who was driving on her own pulled into a filling station and bought some gasoline. As she drove off she noticed a stranger in a car following her. She tried to shake him off by turning, accelerating, slowing down, etc. Finally she turned into a police station, but she was shocked to see him follow her in. He was not a policeman and there was no mechanical faults with her car. Why did he follow her? (Sloan and MacHale 1994, p. 18.)

The solution turns on the introduction of a new individual:

He had seen a man hide in the back of the woman's car as she paid at the gasoline station. He followed her to warn her and was pleased to see her pull into the police station. (Ibid., p. 80.)

REFERENCES

Adkins, Arthur, 1960, *Merit and Responsibility*, Clarendon Press, Oxford.

Barnes, Jonathan, translator and editor, 1994, *Aristotle's Posterior Analytic*, second ed. (Clarendon Aristotle Series), Clarendon Press, Oxford.

Beth, E.W., 1955, "Semantic Entailment and Formal Provability," *Mededelingen van de Koninklijke Nederlandse Akademic van Wetenschappen, Afd. Letterkunde*, N.R. vol. 18, no. 13, Amsterdam, pp. 309-342.

Cornford, F.M., translator and editor, 1945, *The Republic of Plato*, Oxford University Press, New York.

Eatwell, John, Murray Milgate and Peter Newman, editors, 1989, *The New Palgrave: Game Theory*, W.W. Norton, New York.

Gärdenfors, Peter, 1988, *Knowledge in Flux*, MIT Press, Cambridge.

Gentzen, Gerhard, 1969 (edited by M.E. Szabo), *The Collected Papers of Gerhard Gentzen*, North-Holland, Amsterdam.

Henkin, Leon, 1949, "The Completeness of the First-order Functional Calculus," *Journal of Symbolic Logic* vol. 14, pp. 159-166.

Hintikka, Jaakko, 1996, "On the Development of Aristotle's Ideas of Scientific Method and the Structure of Science," in W. Wians, editor, *Aristotle's Philosophical Development: Problems and Prospects*, Rowman & Littlefield, Savage, Maryland, pp. 83-104.

Hintikka, Jaakko, 1989, "The Role of Logic in Argumentation," *The Monist* vol. 72, no. 1, pp. 3-24.

Hintikka, Jaakko, 1987, "The Fallacy of Fallacies," *Argumentation* vol. 1, pp. 211-238.

Hintikka, Jaakko, Ilpo Halonen, and Arto Mutanen, forthcoming, "Interrogative Logic as a General Theory of Reasoning," in Ralph H. Johnson and John Woods, editors, *Handbook of Practical Reasoning,* Kluwer Academic, Dordrecht.

Padolfini, Bruce, 1986, *Kasparov's Winning chess Tactics*, Simon and Schuster, New York.

Plantinga, Alvin, 1993, *Warrant: The Current Debate*, Oxford University Press, New York.

Salmon, Merrilee H., 1989, *Introduction to Logic and Critical Thinking*, second ed., Harcourt, Brace, Jovanovich, San Diego.

Sloan, Paul and Des MacHale, 1994, *Great Lateral Thinking Puzzles*, Sterling, New York.

Smullyan, R.M., 1968, *First-order Logic,* Springer-Verlag, Berlin-Heidelberg-New York.

Suppes, Patrick, 1957, *Introduction to Logic,* Van Nostrand, Princeton.

Toulmin, Stephen, R. Ricke and A. Janik, 1979, *An Introduction to Reasoning,* Macmillan, New York.

von Neumann, John and Oskar Morgenstern, 1944, *Theory of Games and Economic Behavior*, Princeton University Press, Princeton.

Wang, Hao, 1993, *Popular Lectures on Mathematical Logic*, Dover Publications, New York.

APPENDIX: A CASE STUDY

As an example of the relationships between verbal, pictorial and formal reasoning, let us consider the following easy puzzle:

> A man and his sister were out shopping one day when the man said, "That boy over there is my nephew." "That is right," replied his sister, "but he is not my nephew." How come? (Sloan and MacHale 1992, p. 12.)

One natural line of reasoning that can be used to solve the puzzle (expounded somewhat pedantically) runs as follows:

For the boy to be the man's nephew means that (1) there is someone who is the man's sibling and the boy's parent. (2) Call that person, for the sake of argument, Pat. (3) Pat is the man's sibling, as (4) is the sister. (5) Pat is also that boy's parent. But (6) if two different people are the same person's siblings they are each other's siblings. Hence (7) if Pat and the sister are different persons, they are siblings. But the boy's not being the sister's nephew means that (8) none of the sister's siblings is one of the boy's parents. Accordingly, (9) if Pat is the sister's sibling, he or she is not the boy's parent. But (5) Pat *is* the boy's parent, wherefore (10) Pat is not the sister's sibling. But, (11) Pat would be the sister's sibling if he or she were different from the sister. Hence (12) Pat is the sister and (13) the boy's mother.

This argument is very intuitive. It can even be illustrated by a figure, perhaps something like this:

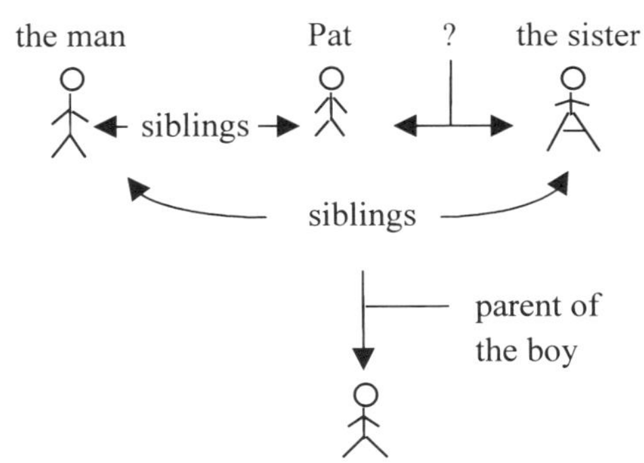

The same intuitive argument can be represented by means of the tree method or *tableau* method (left side only) by means of an obvious key (m = the man, s = his sister, β = Pat, Sxy = x is a sibling of y, Pxy = x is a parent of y) as follows:

(1) (∃x)(Sxm & Pxb) (A)

(2) (Sβm & Pβb)
from (1)

(3) Sβm
from (2)

(4) Ssm (A)

(5) Pβb
from (2)

(6) (∀x)(∀y)(∀z) (A)
(((x ≠ y) & Sxz & Syz) ⊃ Sxy)

(7) (β ≠ s) & Sβm & Ssm) ⊃ Sβs)
from (6)

(8) (∀x)(Sxs ⊃ ~Pxb) (A)

(9) Sβs ⊃ ~Pβb)
from (8)

(10) ~Sβs
from (9), (5)

(11) (β ≠ s) ⊃ Sβs)
from (3), (4), (7)

(12) β = s
from (10), (11)

(13) Psb
from (12), (5)

Here (A) indicates an assumption of the puzzle.

Several observations can be made here. First, the verbal argument and the formal one match each other perfectly. If one is careful in spelling out one's assumptions one can construct as strict an argument verbally as one can do formally. Needless to say, in the formal version it is much easier to ascertain that all the assumptions are correctly expressed and that all steps of the argument are logically valid.

Furthermore, the equivalence of the formal and the intuitive argument illustrates the model-theoretical interpretation of first-order arguments. What either argument is, is literally an attempt to describe what a scenario would be like in which all the premises of the argument are true. This applies to the formal argument quite as much as to the informal one.

It can also be seen that both the formal and the informal argument depend essentially on focussing who is the boy's parent and hence makes the boy the man's nephew. What is his or her relation to the sister? In the formal argument, that person is introduced by the instantiation step (2). Intuitively, it means introducing a new individual into the reasoning. It is precisely this need of introducing new individuals or more generally speaking new functions into an argument that makes it nontrivial. This general truth is not spoiled by the fact that in this particular example the unknown new individual (Pat or β) eventually turns out to be identical with one of the old principals. This illustrates the strategic importance of appropriate introductions of new individuals and functions.

In other words, the crucial step in this line of thought is the introduction of the "arbitrary individual" Pat. Without this introduction, under some guise or another, the argument would not succeed. We can also see that this is an exceptionally simple case in two different respects. First, we have little choice as to which new individuals to introduce. In general, the reasoner faces a choice of many different possibilities of existential instantiations (on the left side). Then, as was emphasized in the paper, the choice of the right instantiations (of the right new individuals to introduce) will be the crucial strategic choice.

It can also be seen that many of the propositional logic steps are easily, perhaps even normally, glossed over in an informal argument. For instance, in the line of thought above some of the steps (e.g. steps like (7), (10)-(12)) are likely to remain tacit in an informal argument. Logic teacher's insistence on not omitting such intermediate steps is not conducive of developing efficient reasoning methods in their students. Conversely, the verbal argument appears easier than the formal one to construct only insofar as the reasoner skips some of the more obvious steps. If the verbal argument is to be strict, it will have to be as elaborate, not to say tedious, as the formal one.

At the same time, it is obvious that even though the formal argument merely reproduces the intuitive verbal one, it does not make the argument any easier to discover. No matter whether you represent the argument formally or verbally, the thought process by means of which you solved the original puzzle is likely to be the same kind of thought-experiment as was carried out in the intuitive reasoning above. You imagine what it would mean for the boy to be a nephew of the man but not his sister. Thus of the three equivalent ways of looking at the argument, the one is heuristically the superior one.

Moreover, formalization taken alone is not really helpful for the purpose of mechanical proof construction, either. The reasons are largely combinatorial. Let us see first what happens when only cut-free rules are used. In our example, the general conditional (6) could have been instantiated universally in 64 different ways. Yet only one of them was needed for the argument. (The others would have been totally unhelpful.) A computer program without a strategy of paring down the set of possible construction lines would have been forced to examine hundreds of formulas in trying to find the proof. And to try to find a single line of argument would have presupposed knowing how to reach it, instead of letting the computer discover the strategy.

In this case, a computer could have reached the proof by simply examining all possible construction lines. But in ever so slightly more complex cases, the sheer number of formulas would have quickly become impossible to handle even for the most powerful computers.

2

THE ROLE OF LOGIC IN ARGUMENTATION

1. The Sherlock Holmes sense of logic and its decline

The main currently unsolved problem in the theory of argumentation concerns the function of logic in argumentation and reasoning. The traditional view simply identified logic with the theory of reasoning. This view is still being echoed in older textbooks of formal logic. In a different variant, the same view is even codified in the ordinary usage of words such as 'logic', 'deduction', 'inference', etc. For each actual occurrence of these terms in textbooks of formal logic, there are hundreds of uses of the same idioms to describe the feats of real or fictional detectives. I have called the idea reflected by this usage the "Sherlock Holmes conception of logic and deduction." In the history of science, we find no less a thinker than Sir Isaac Newton describing his experimental method as one of analysis or resolution and claiming to have "deduced" at least some of his laws from the "phenomena."[1]

In fact, one way of approaching the problem of the role of logic in argumentation is to try to discuss and to evaluate this traditional view of the nature of logic.

Needless to say, this traditional conception of logic and deduction has been rejected with a rare unanimity by recent theorists of human reasoning and argumentation. It is widely assumed that the truths of formal logic are mere tautologies or analytical truths without substantial content and hence incapable of sustaining any inferences leading to new and even surprising discoveries, as the detections of sleuths like Sherlock Holmes or Nero Wolfe were supposed to lead. Newton's claim of having "deduced" his laws from phenomena is currently dismissed by historians and philosophers of science as mere window-dressing, without any real connection with his scientific practice—when not turned into an object of scorn by such opinionated philosophers as Karl Popper. Some scholars have looked for tools of argumentation in traditional theories of rhetoric rather than logic.

2. There is no theory of informal reasoning

But all this criticism of logic as an *Organon* leaves unanswered the question of the actual role of logic in reasoning and argumentation. This is the unanswered question I mentioned. For one thing, there surely is something to the old idea of logic as the organon of reasoning in general. Whom do you believe when it comes to actual nitty-gritty scientific reasoning, logic-oriented philosophers of science or rhetoricians? Can half-a-billion readers of detective stories be completely wrong? Has any scientist ever been more thoughtful about his own methodological views than Isaac Newton? Surely the rhetorical evidence tells here against the idea of argumentation as an exercise in mere non-logical rhetoric.

This problem is an especially burning one on the pedagogical level. Philosophers assigned to teach introductory courses in formal logic have often been dissatisfied with the educational value of what they have conveyed to their students. As a consequence, not only has a plethora of courses in"reasoning and criticial thinking" or "argumentation theory" sprung up, but there even exists an organized "informal logic" movement dedicated to developing a theory and a practice of informal argumentation.

Unfortunately, there exists by any reasonable standard no respectable general theory of informal argumentation. It serves no constructive purpose to argue in detail for this judgment here. If your intellectual tastes are such that you are satisfied with the theoretical level of traditional rhetorical theories or of so-called theories of informal reasoning, I am not going to be able to re-educate you in half-an-hour.

Let me nevertheless mention, as an example, one of the ideas that has recently gained some currency. It is the idea that actual reasoning should not be approached as if it were a chain of deductive inferences. Rather, what the theory of substantial inference is, is the body of principles of belief-changes in the presence of new evidence.

As a heuristic idea, there is little to object to in the approach just mentioned. Yet it has scarcely led to anything remotely like a satisfactory theory of reasoning in general. The following are among the most glaring weaknesses: (1) No theory has been developed as to where the new evidence itself is to be found. Nor is this new approach capable of handling questions of reasoning strategies in any other size, shape or form. (2) It does not present any explanation of the true element in the traditional conception of logic as a general theory of reasoning. (3) This type of approach often relies on notions like "inference to the best explanation." Such notions seem to be either too vague, too complicated, or too little understood, to sustain as yet a genuine theory of the subject.

3. *Definitory* vs. *strategic rules*

A partial diagnosis of the situation turns on a misunderstanding which has haunted the discussion of such ideas as "rule of inference." Here a quick side-glance at the theory of games serves to clear the air. In games, there are rules and there are rules. There are such rules as serve to define the game, e.g., the rules of chess. I shall call them 'definitory rules'. They tell which moves are possible, or, as it is sometimes put, which moves are admissible. The crucial fact about definitory rules is that they say absolutely nothing about which moves are good, which ones are bad, and which ones are better than others. Such questions are handled by rules of another kind. I shall call them 'strategic rules'. They have to be distinguished from definitory rules. Admittedly, the notion of strategy in a given game is possible to define only after the definitory rules have been set up. Only after that has been done can we hope to begin to investigate which strategies are better than others.

When we do so, we soon realize that particular moves cannot always be compared with others for their merit. For here we come to the other general insight which game theory helps us to reach. It is that strategic rules apply in principle only to, well, strategies. Absolutely speaking, we cannot always say that one move is better than another move, for that depends on the overall strategies as members of which the moves in question are being considered. In this respect, strategic rules differ fundamentally from definitory rules, which typically deal with the inadmissibility of particular moves. For, in game theory, values (a.k.a. utilities or payoffs) can in the last analysis be associated only with strategies. Particular moves receive their "value" from the strategies as a part of which they are considered.

This fundamental contrast is not changed nor is its importance diminished by a meta-observation concerning the methodology of game-model construction. Often, a meta-theorist who is trying to construct a game-theoretical model for certain phenomena has a choice. The meta-theoretist has normally different games to choose from by means of which a certain process is conceptualized. It may happen that the same behavior on the part of one of the players is in one game a breach of rules, but in another merely a bad (costly) move. In other words, what in the former game is a breach of a definitory rule is in the latter one a breach of a strategic rule. (Thus Talleyrand was construing politics as a much more sinister "game" than his interlocutor when he said of the execution of the duc d'Enghien: "It's worse than a crime, it's a mistake.") This freedom on the part of a meta-theorist does not in the least diminish the importance of the original distinction. On the contrary, the meta-theorist's option cannot be appreciated without understanding the distinction between definitory and strategic rules.

4. Strategic rules have been neglected in deductive logic

Now in the "games" of formal logic the rules of inference act as definitory rules. They say absolutely nothing about good inferences or bad ones. A rule of inference does not even recommend the inferences it allows. It merely serves to delineate inferences from non-inferences, moves from non-moves, not good moves from bad ones. Yet it is precisely rules of inference that receive a lion's share of time and attention in the usual courses of formal logic. This is, of course, precisely the wrong way around if logic is to be thought of as the science of reasoning. Nobody will become even a mediocre chessplayer merely by learning which moves of the different chessmen are admissible. One becomes a chessplayer worthy of the term only when one also learns a modicum of strategic rules. (Of course, the more elementary ones should be called tactical rather than strategic rules.) And the same is true of the "game" of deductive logic.

Thus, in a sense, the same fate has befallen the study of logic as the study of ethics. Once upon a time, ethics was primarily the study of moral excellence, of the different virtues and of the ways of achieving them. A virtuous person was literally a moral virtuoso. Alas, a funny (or sad) thing happened on the way to Victorian and post-Victorian morality. In the prevailing popular view, a virtuous person was one who merely avoided moral mistakes; on an extreme view, a person who did not lose his or her virtue, as the Victorian euphemism went. And this popular view, codified as it was in the ordinary semantical force of expression such as "virtue" and "virtuous", and in the so-called "intuitions" of men and women in the street, cannot but have influenced philosophers' ethical theories, based as they are in our day on our alleged "intuitions" about moral matters.

It is no exaggeration to say that the ethos that pervades many introductory textbooks of logic is a theoretical analogue to ethics as a study of ways of avoiding moral lapses. For the concentration of attention on the definitory rules called rules of inference means that the main attention is directed at avoiding mistakes, avoiding steps that are ruled out as not qualifying as inferences at all, neither as good ones nor as bad ones. Avoiding logical sins becomes, even if the textbook writer did not intend it, more important than excellence as a logician. Preserving one's logical virtue becomes a more important concern than developing virtuosity in drawing logical inferences.

Admittedly, definitory rules must be learned first before strategic rules can be as much as understood. Moreover, beyond the most trivial parts of logic, there will not be any completely general, mechanically applicable (i.e., recursive) rules defining optimal strategies. But noting these facts of a logician's life is a poor excuse for the neglect of the entire strategic viewpoint, which we find in most textbooks of formal logic (and, for that matter, in many textbooks of informal logic, too). Even if strategy is an art and not a science when it comes to logic, it should be taught and learned.

Furthermore, the rules of inference, be they formal rules of deductive inference or rules for "inference to the best explanation," cannot be the proper tools for studying skillful inferences. For skill in reasoning is a matter of strategy (and tactics). And, as we saw, matters of strategy choice cannot be discussed move by move. They pertain to the entire organization and ordering of particular steps of inference. They cannot be adequately discussed by reference to (definitory) rules of inference, for such rules traffic in particular moves in the "game" of deductive logic. Thus, the neglect of strategic considerations on the part of logic teachers and textbook writers has not been a matter of biased attention only. It is partly due to, and manifests itself in, the use of inappropriate conceptual tools.

5. *Corollarial* vs. *theorematic inference*

One prominent victim of this one-sided orientation of logical theory has been the distinction I made some time ago between what I called trivial and non-trivial steps of reasoning.[2] It has turned out since that my distinction had been strikingly anticipated by Charles S. Peirce who used the more telling terms "corollarial" (for trivial) and "theorematic" (for non-trivial) reasoning.[3] The only real difference is that I showed explicitly how to draw the distinction also in a completely formalized first-order (quantificational) reasoning. It is a sad testimony to philosophers' biased orientation towards logic that this explication of the distinction should have prompted the accusation that I am turning Peirce's distinction into a merely logical matter. I shudder to think what scorn Peirce would have poured on the assumption that distinctions made in purely logical terms cannot be relevant to, or cannot serve to clarify, substantial matters, such as the nature of reasoning in general.

Indeed, if philosophers had been interested in matters of deductive strategy and tactics, they would have had every reason to welcome Peirce's distinction. For even though it deals in the first place with particular steps of argument, it shows which kinds of deductive inference can, and which ones cannot, open a door for further argumentative possibilities. The basic intuitive idea is almost ridiculously simple. In a sense that can be made precise, theorematic inferences are the ones which introduce a new individual into the argument, whereas corollarial inferences merely traffic in the individuals which have already been considered in their premises. From this intuitive description, it is seen at once how theorematic inferences can open new lines of thought for one's argumentation (even if they don't automatically do so). Hence Peirce's distinction is an important prolegomenon for any serious study of deductive strategy. And yet it fell stillborn from Peirce's pen, and the same fate has more recently befallen its born-again version. This neglect is a telling example of philosophers' disinterest in questions of deductive strategy.

But even if this disinterest in argumentative strategy in the case of purely deductive argumentation helps to explain logic teachers' disenchantment with current textbooks of symbolic logic, it does not yet answer the paramount question of the relation of logic to argumentation in general.

Before essaying such an answer, a few supplementary comments may be in order. The distinction between theorematic and corollarial deductive reasoning can be developed further and turned into a difference in degree rather than of kind. The way of doing so is obvious on the basis of what has been said. If theorematic inferences are characterized by the introduction of new individuals into the arguments, then the total number of such new individuals introduced during a line of thought will serve as an index of its theorematicity.

This observation can once again be turned against my poor *bête noire*, the average logic-textbook writer. For even when theorematic deductive arguments are considered in logic texts, the total number of new individuals introduced in them (i.e., the index of theorematicity of such arguments) is always quite small. In reality, it is known that the index of theorematicity of a deductive argument from a premise T to a conclusion C, considered as a function of the Gödel numbers of T and C, grows eventually faster than any recursive function.[4] In other words, typically this index is astronomically large. Of course, "typically" does not mean here "in cases most often met with in practice," but rather "in cases illustrating the theoretical situation." But even so, one might reasonably hope that logic texts would give their readers a sense of this overall theoretical situation.

6. The interrogative model of inquiry as a general theory of reasoning

In this paper, I shall outline a solution to the problem of the relation of deductive logic to the theory of reasoning and argumentation in general. This solution is formulated in terms of what I have called the interrogative model of inquiry. In this paper, I cannot argue adequately for the status of this model (this kind of model) as a codification of human reasoning in general, beyond pointing out how neatly my epistemology reproduces etymology here. What I am doing is merely to conceptualize inquiry in the sense of investigation, or search for truth, as inquiry in the sense of a process of questioning.

Persuasive evidence is perhaps obtainable by considering what my overall claim implies. Among other things, I am implicitly claiming that all the clever "deductions" of real or fictional detectives can be represented as question-answer sequences. For instance, I must be able to rewrite the solution of any Sherlock Holmes story in an interrogative form. The example I have used most often is one of the best known Sherlock Holmes episodes, "the curious incident of the dog in the night-time." It is from the story called "Silver Blaze." The background is this: the famous racing-horse Silver Blaze has been stolen from the stables in the middle of the night, and in the morning its trainer, the stablemaster, is found dead out in the heath, killed by a mighty blow. All sorts of suspects crop up, but everybody is very much in the dark as to what really happened during the fateful night until the good inspector asks Holmes:

"Is there any point to which you would wish to draw my attention?"

"To the curious incident of the dog in the night-time."

"The dog did nothing in the night-time."

"That was the curious incident," remarked Sherlock Holmes.

Even Dr. Watson can see that Holmes is in effect asking three questions. Was there a watchdog in the stables when the horse disappeared? Yes, we have been told that there was. Did the dog bark when the horse was stolen? No, it did not even wake the stable-boys in the loft. ("That was the curious incident.") Now who is it that a trained watchdog does not bark at in the middle of the night? His owner, the stable-master, of course. Hence it was the stable-master himself who stole the horse . . . . Elementary, my dear Watson.

Such successes in reconstructing interesting non-logical "deductions" as question-answer sequences offer cumulative evidence for the applicability of the interrogative model.

7. *The structure of the interrogative model*

The basic "Model T" version of the interrogative model is simplicity itself. It is naturally (i.e., in view of the importance of the strategic viewpoint) formulated in the form of a game between an idealized *inquirer* and a source of answers which will be referred to as *nature* or the *oracle*. In different applications of the model, this "oracle" will be quite different. For instance, observations can be thought of as answers put to one's environment. To some extent, this may be true of perception in general. Psychologists such as von Helmholtz have emphasized the role of what they call "unconscious inferences" in perception. Such "inferences," it is only fair to surmise, fall within the scope of my attempt to treat inferences and reasoning in general as question-answer sequences. Hence von Helmholtz might have spoken about unconscious questions rather than about unconscious inferences.

In other applications, the question put to nature can be a controlled experiment. In still other cases the "oracle" can be the database stored in the memory of a computer, a diagnostic handbook, a witness in a court of law, or one's own tacit knowledge partly based on one's memory.

I am not excluding the possibility that there may be interesting structural (logical) differences between these several types of applications, for instance with respect to the quantificational hierarchy outlined below.

In all these cases, the inquirer may be thought of as starting from a given theoretical premise T. In one type of interrogative process, the aim of the game is to establish a certain given consequence C. In another type of game, the inquirer is trying to prove either B or not-B, that is, answer the question "B or not-B?" (I have proposed to call this the Shakespearean variant of the questioning games.) In others, the goal is to answer some more complicated question.

For each particular conclusion, the process looks on paper rather like any old deduction. I shall concentrate in what follows on the given-conclusion case, and use as the book-keeping device a Beth-type semantical *tableau*.[5] (Such a *tableau* is of course but a Gentzen-type sequent calculus deduction turned upside down.)

The deductive moves are governed by the usual *tableau* construction rules, suitably modified so as to avoid unwanted assumptions. Among other restrictions, the sub-formula principle must be adhered to strictly, and all traffic from the right column to the left must be curtailed.

Apart from such technical restrictions, my *tableaux* will look rather like Beth's, except for one thing.[6] At each stage of the game, instead of making a deductive move of *tableau*-construction, the inquiry may address a question to the answerer (oracle, nature, or whatever the source of new information may be). If nature responds, the answer is entered in the left column as an additional premise. Such a move will be called, unsurprisingly, an interrogative move.

8. Interrogative moves and restrictions on answers

These interrogative moves will need a number of further comments.

(i) The interrogative model may appear so simple that it will not offer any interesting novelties. This impression is mistaken for an interesting reason. The interrogative model is not well defined before the precise relation of questions to the (conclusive) answers and to their presuppositions is mastered. Such mastery has been achieved only relatively recently.

(ii) Here the presuppositional relation is the crucial one. In the simplest case, we are dealing with two kinds of questions, propositional questions and simple wh-questions.

The presupposition of the former is of the form

(1) $(S_1 \vee S_2 \vee \ldots \vee S_k)$

and its conclusive answer is

(2) S_i $(i=1, 2, \ldots, \text{or } k)$.

The presupposition of a simple wh-question is

(3) $(\exists x)\, S[x]$

and its conclusive answer is of the form

(4) $S[b]$

where "b" is the name of the individual in the model (system, "world") by reference to which the interrogative process takes place.

(iii) An important part of the full interrogative game is the specification of the answers which the oracle will actually give to the inquirer's questions. This specification can be made in a large variety of different ways. Sometimes it is implicit in the defining conditions of an application For instance, the answers one can extract from a database are simply the data stored therein.

It turns out that one particularly interesting dimension of restrictions on the oracle's answers is according to the quantificational complexity of available answers. In the simplest (atomistic) case, nature (the oracle) gives only quantifier-free answers. In the next one, the A-case, nature will give answers whose quantifier prefix consists of universal quantifiers only. In the next one the quantifier prefixes have the form

(5) $(\forall x_1) \ldots (\forall x_i)\, (\exists y_1) \ldots (\exists y_j)$

This will be called the AE-case. And so the hierarchy continues, with the (quantificationally) unrestricted case as the other extreme.

An especially important question is: Which case along the quantificational-complexity dimension characterizes which type of application? I have discussed some special cases of this general problem elsewhere, with interesting results.[7]

It can, for instance, be argued that the "deductions" of clever detectives belong to the unrestricted case. Sherlock Holmes may, so to speak, evoke from his extensive tacit knowledge of such sciences as chemistry any old law of arbitrary quantificational complexity, whatever other limitations there may be to the answers available to the great sleuth.

9. *Definitory* vs. *strategic rules again*

The same distinction between definitory rules and strategic rules as was discussed above in connection with deduction applies to interrogative games with a vengeance. In other words, one of the main new types of studies which the interrogative model opens for us is to strategies of questioning, that is, strategies of information seeking by means of the different choices of questions to be asked and of the order in which they are asked. It is not much of an exaggeration to say that here we have the most important new opportunity which the interrogative model facilitates.

As a beginning step in this direction, it may be pointed out that the distinction between corollarial and theorematic reasoning can clearly be extended to the entire interrogative inquiry, and not only to its deductive component. For there are two identifiable ways, and two ways only, in which new individuals can be introduced into an interrogative argument. As in the purely deductive case, new individuals may be introduced by existential instantations as values of instantiating terms, sometimes called "dummy names." Values of such dummy names are of course not real individuals, only "arbitrary objects" or "random individuals." Yet they add in a real sense to the complexity of the configurations of individuals considered in an argument.[8]

In an even more concrete sense, new individuals may be introduced by the oracle's answers to wh-questions. In this case, the new individuals are actual members of the domain of individuals of the model ("universe of discourse") in relation to which the interrogative game is played.

The distinction between corollarial and theorematic reasoning allows in fact striking illustrations in its generalized form. Consider, for instance, "the curious incident of the dog in the night-time" examined above in sec. 6. Up to Sherlock Holmes's remark, everybody has been considering only the principals of the case: the stolen horse, its unknown thief, the dead stable-master, and his killer. (Of course we do not know whether they are actually different from each other.) What Holmes is doing is to introduce a new entity into the consideration, viz., the dog. This is precisely the kind of introduction of a new individual (in a logician's sense) into an argument that is needed. And, lo and behold, some light is at once thrown on the events of the mysterious night-time. This is in fact a good example of how the introduction of new individuals can help one's reasoning.

10. Division of labor between deductive and interrogative moves

At this stage of my line of thought, the relation of logic (deduction) to reasoning in general, conceptualized by means of the interrogative model, allows for a preliminary formulation. The relationship in question is one of a division of labor. Deduction (logic) and interrogation appear as two interacting and mutually reinforcing components of inquiry. Neither is dispensable. Questions are needed to bring in substantially new information, and deductions are needed both for the purpose of spelling out the consequences of such information and, more importantly, for the purpose of paving the way for new questions by establishing their presuppositions. Thus one important source of the need of the deductive component in inquiry is the role of presuppositions in questioning. This is one of the many ways in which the study of the logic of questions can help us to understand the nature of inquiry in general. Small wonder, therefore, that such thinkers as Collingwood and Gadamer have seen in the logic of questions an important aspect of the nature of inquiry in general, and small wonder that they paid special attention to the presuppositions of questions.[9]

The interactive relationship between deductions and questions deserves a few further comments. First, if the interrogative model is a half-way natural representation of at least some important types of non-logical reasoning (argumentation), logic (of the kind usually codified in formal deductive logic) is an indispensable component of reasoning in general. There is no such thing as completely informal logic of argumentation or reasoning. The very term "informal logic" is a solecism.

Furthermore, there is no absolute sense in which one of the two intertwined components of interrogative inquiry, deductions and questioning, is more important or more difficult, absolutely speaking. Such judgments can only be made on the basis of some particular assumptions concerning the "cost" of different kinds of moves in the interrogative games of inquiry. A game theorist would codify such assumptions in the "payoffs" of the game. There is no unqualified reason why an interrogative inquiry could not contain such long chains of non-trivial ("theorematic") deductions as were discussed earlier.

It is also important to realize that the role that logic plays in the overall process of inquiry in the form of deductive moves cannot usually be separated from the rest. Deductive "moves" are needed in the first place for the purpose of establishing presuppositions for interrogative moves. Thus the order of the two kinds of moves cannot in general be changed so as to segregate them from each other. Only in those special cases in which presuppositions of questions are dispensable or trival can we hope to rearrange the applications of the two kinds of rules, e.g., in such a way that all questions precede all deductions. The most likely case for this rule of the atomistic case.

11. Origin of deductive logic

At the same time, the interrogative model and its history helps to uncover certain contingent reasons why the role of deductions in inquiry easily appears unimportant. First, genetically speaking formal logic seems to have come about in the context of a study of the questioning games which were played in Plato's Academy, partly inspired by the Socratic method of questioning.[10] In them, there were no deductive steps as a distinguishing class of "moves." What apparently happened was that Aristotle perceived that the answers to suitable questions were prejudged by the answerer's earlier responses or "admissions." No rational person could answer these questions in more than one predictable way. Hence recognizing such questions and answers is an especially useful tactical skill in the Platonic questioning games.

Such predetermined answers are precisely the ones which are logically implied by earlier ones. Thus the study of formal deductive logic arose from Aristotle's study of the strategy and tactics of interrogative argumentation.

But it thus arose in a way which encouraged philosophers to underestimate the deductive component of argumentation in general. For even though all deductive consequences of the answerer's earlier admissions are in a sense predetermined by his earlier replies, this predetermination is not always obvious. They may in practice have to be elicited from the answerer by means of further questions. Only in the case of corollarial inferences are logical consequences of one's opponent's earlier answers even arguably obvious. But Aristotle never acknowledged in his logical theory any non-trivial (theorematic) logical inferences nor managed to cope in his logic with the theorematic inferences which he was familiar with from geometry.[11] Thus, there was from the very inception of formal logic associated with it a mistaken stigma of triviality. There are important further reasons for the fallacious reputation of formal logic for triviality. I will return to them in sec. 17.

12. A parallelism between deductive and interrogative moves

All this nevertheless does not yet touch the most important aspect of the relationship of deductive logic to reasoning in general. This aspect begins to unfold for us when we note certain important facts about the relationship of deductive moves to interrogative moves in the "games" of inquiry outlined above.

Suppose that there occurs in the left column of a game *tableau* (or *subtableau*) an existential sentence

(6) $(\exists x)\ S[x]$.

What this situation amounts to is of course one in which the inquirer has either proved that $(\exists x)\ S[x]$ is true or assumed it to be true. How can the inquirer try to exploit such a situation? In two different ways.

(i) The inquirer can apply the deductive (*tableau* construction) rule of existential instantiation to the sentence in question. The output (formula added to the left column) is of the form

(7) $S[\alpha]$.

Here "α" is not a name of any actual member of the universe of discourse, but a free variable. It is sometimes called, aptly, a "dummy name." If it is thought of as standing for something, than something has to be thought of as an "arbitrary individual"—whatever such an animal is or may be.

A concrete consequence of this status of "α" is that the inquirer cannot ask any questions in which the term "α" occurs. For clearly an "arbitrary object" does not have any definite properties that one could inquire about.

A completely satisfactory interpretation of the role of "dummy names" can only be obtained by means of the original idea on which a *tableau* construction is based. Such a construction means an attempted construction of a counter-model, i.e., a model in which the premise T would be true but in which the hoped-for conclusion C would not be true. Even though "dummy names" like "α" stand for individuals in the attempted model, they do not stand for objects in the real world.

Speaking of "arbitrary objects" may be heuristically suggestive, but it will not yield philosophically viable understanding of the situation.

(ii) But the inquirer can try to exploit the presence of (6) in the left column in another way. The inquirer can use (6) as a presupposition of a wh-question. Since the presupposition has been established, the inquirer may ask the question. Whether an answer is forthcoming depends on the restrictions that have been placed on the oracle's responses.

Let us assume, for the sake of argument, that nature (the oracle) answers the question. The answer will be of the form

(8) S[b]

where "b" is a proper name of an element in the domain of individuals of the "world" in relation to which the interrogative game takes place.

The two outcomes (7) and (8) are similar in structure. How do they compare with each other as far as the rest of the interrogative game is concerned? Everything the inquirer can do by means of (7) he or she can do by means of (8). But not vice versa, for the inquirer may very well be able to use (8) as a basis for further questions, whereas (7) was seen to be incapable of serving in that role.

Hence, when it comes to a choice between the deductive move (i) and the analogous interrogative move (ii), the inquirer is well advised to ask the question instead of making a deductive move, provided that the oracle will answer the question. At the very most the inquirer in doing as well if he asks the question (and obtains an answer), and normally he or she is doing better in the rest of the interrogative argument on the basis of (8) than on the basis of (7).

Similar observations can be made when a disjunction

(9) $(S_1 \vee S_2 \vee \ldots \vee S_k)$

is present in the left column of a *(sub)tableau*. Again, the inquirer has two options.

(i) The inquirer may apply the relevant deductive rule. That results in splitting the *(sub)tableau* construction into k *(sub)subtableaux*, each of which has S_i (i=1, 2, . . . , k) as its sole new formula in the left column.

(ii) The inquirer may use (9) as the presupposition of a propositional question. If the oracle answers this question, the answer is of the form S_i (i=1, 2, . . . , or k).

Which course of action is more advantageous to the inquirer? If the oracle ("nature") actually responds with an answer, the inquirer is better off with this answer. For then he or she has to try to close only one of the k different *subtableaux* which the inquirer had to worry about in (i).

Once again, it is therefore never advantageous for the inquirer to make a deductive move if he or she can ask a parallel question and have it answered.

Furthermore, the two types of deductive moves, existential instantiation and the splitting of a tableau construction into several, are the only non-trivial deductive moves, as far as the left column is concerned. The other rules (the universal instantiation rule and the conjunction rule) can only be applied a finite number of times in any one situation. They can never open up essentially new opportunities of rule application or split up the attempted countermodel construction into several.

13. Sherlock Holmes's sense of logic vindicated

These simple observations open an extremely important perspective on the role of logic and of the rules of logic in the general theory of reasoning. As was noted above, at first sight the interrogative model assigns to logic a status of "separate but equal" in relation to the rest of interrogative reasoning. It might even appear possible to study the other facets of reasoning independently of the principles of formal logic. What we have seen is that logic is integrated into the mainstream of inference much more firmly than that.

The more superficial consequence can be formulated as a vindication of the old usage in which "logic" is thought of as a label of any old inference. I have called this usage the Sherlock Holmes sense of 'logic'. A similar usage is found with words like 'deduction', 'inference', etc. This vindication can be explained as follows:

What we have seen means essentially that, in so far as answers are forthcoming to the inquirer's questions, it is advantageous for him or her to ask a question instead of performing the corresponding deductive move. Thus question-answer pairs can do the same job as deductive inferences, normally even better (provided that an answer is actually available). Hence it is eminently natural to call the question-answer step, too, a "deduction" or "inference." It serves the same purpose, after all, as a deductive step in logicians' narrow sense of 'deduction'. But this is precisely the usage I have called the Sherlock Holmes sense of 'deduction', 'inference', 'logic', etc. Far from being an equivocation on these words, it can now be seen to be but a slight extension of the use of these terms in deductive logic. What could be a more complete vindication of Sherlock Holmes?

It is to be remembered, of course, that the parallelism between deduction and interrogation holds in so far as the inquirer's questions are actually answerable. In so far as the quantificational hierarchy mentioned above in sec. 8 is concerned, the type of inquiry that comes closest to automatic answerability is the unrestricted case. This is very much in keeping with our line of thought here. For it was also pointed out above (in sec. 8) that the type of inquiry instantiated by the famous "deductions" of the likes of Sherlock Holmes is precisely the unrestricted case. This makes the extenstion of terms like 'deduction', 'inference', 'logic', etc. especially natural in their mouth.

14. A strategic parallelism between interrogation and deduction

The most important observation concerning the role of logic in interrogative argumentation nevertheless still remains to be made. In deductive reasoning, it is known that one of the crucial determinants of good strategy is the choice of the right existential formulas (in the left column) or, by the parity of cases, universal formulas (in the right column) to instantiate. This strategic fact of a deductive logician's life is seen, e.g., if you are trying to program a computer to carry out deductive strategies. But even if you are not familiar with the studies of mechanical theorem-proving in which this massive fact emerges especially clearly, the point is easy to understand on the basis of what was said earlier in this paper. For it is precisely existential instantiations that introduce new individuals into deductive reasoning and thereby make the total arguments theorematic (non-trivial). Hence the proper choice of these instantiating steps is obviously of a crucial strategic importance in deductive reasoning.

One historically important example of this fact is the deductive argumentation used in elementary geometry. There existential instantiations are known as auxiliary constructions. And perceptive analysts of geometrical reasoning have known since antiquity that the success or failure of geometrical reasoning depends on the right choice of auxiliary constructions.[12]

Now it can be seen that existential instantiations are parallel to answerable wh-questions. Clearly, on the interrogative side, too, it is the choice of which appropriate wh-questions to ask that essentially determines the chances of success of a line of interrogative inquiry. Indeed, it was this very possibility of guiding the course of inquiry by a judicious selection of questions that first prompted Immanuel Kant to compare the experimental reasoning of physical scientists to "putting questions to nature."[13] Thus in interrogative inquiry, too, the selection of the appropriate existential sentences as a basis of the next move is one of the crucial strategic considerations.

Moreover, and most crucially, the general parallelism which we have seen to obtain between interrogation and deduction obviously means that *the strategic principles* governing this choice of the existential sentence to try to instantiate *are* roughly *the same in interrogative inquiry and in deductive logic.*

15. Simplifying extra premises

This conclusion is strengthened by considering the role of tautological extra premises of the form (S v ~ S) in deduction and in interrogative inquiry.

In a purely deductive inquiry such extra premises can always be added to the left column of a semantical *tableau*. One of the first and most basic results of contemporary proof theory is that such tautological extra premises are always dispensable in purely deductive reasoning, if we are merely considering what can or cannot be proved deductively. This is essentially what Gentzen's first *Hauptsatz* says.[14]

However, such tautological premises may not be dispensable when it comes to shortening and otherwise simplifying deductive arguments. In other words, such tautological extra premises can be strategically highly important. In fact, in some ways their introduction in suitable ways is one of the most crucial strategic considerations in deductive reasoning.

Now in interrogative reasoning such tautological extra premises are not always dispensable. The reason is that such disjunctions will enable the inquirer to ask new questions by serving as their presuppositions. Indeed, the introduction of such premises arguably amounts to a widening of the inquirer's range of attention.

Thus the role of tautologies of the form (S v ~ S), added to the left column of a Beth *tableau*, seems to be quite different in purely deductive and in general interrogative arguments. This dissimilarity pertains to the definitory aspects of the "games" in question. As far as strategies of deduction and strategies of interrogation are concerned, the roles of the auxiliary tautological premises are obviously roughly parallel in the two cases. This implies that the strategic principles of choosing the right auxiliary premises in deduction can be used as guides to the choice of the new yes-or-no question to ask in interrogative inquiry. Once again, deduction and interrogative inquiry are closely parallel, when it comes to strategy selection.

16. Deductive logic as a general theory of reasoning strategies

Even though the strategic parallelism between deduction and questioning is not quite complete, it brings to light the true role of logic in reasoning in general. The relation of logic to the rest of inquiry is not of "separate but equal" roles. Or, rather, this is the idea one receives if one only looks at the (definitory) rules of inference in deductive logic and at the formal rules defining how one can put questions to one's source of new information. If we turn our attention to strategic rules instead of definitory rules, we see that the *strategic* rules are largely (albeit not completely) the same in both cases. This means, among other things, that the study of deductive strategy selection largely amounts to a study of the strategy selection in inquiry (reasoning) in general. The (strategic) rules of good deductive reasoning are rules of good reasoning in general.

This result is a striking vindication of the traditional conception of logic as an important ingredient, perhaps as the most important ingredient, in the art of reasoning in general. This fundamental (but partial) identity has been obscured by logicians' and logic textbook writers' concentration on the so-called rules of inference, i.e., on rules definitory of the "game of logic," at the expense of strategic rules of deductive reasoning. This failure was lamented earlier in connection with deductive logic alone. Now we can see that its consequences have extended much more widely than merely deductive logic.

The constructive reaction to the prevalent one-sided attention to rules of inference is not to try to replace formal logic by some mythical "informal logic," but to redirect our attention to where it, in the last analysis, belongs, to the art and science of good reasoning, to the strategies of thinking and argumentation.

Here, then, we have an answer to the question concerning the role of logic in argumentation. Assuming, as I have done here, that the interrogative model is a realistic representation of reasoning in general, we can see that logic plays only a modest supporting role in the total drama of reasoning—on the level of particular steps of inquiry, i.e., on the level of definitory rules. When it comes to questions of strategy, however, logic suddenly assumes a new function. It is as if a supporting actor were at the same time the director of the entire play. On the strategic level, the principles of deductive strategy selection are a major part of the principles of strategy in the entire process of reasoning.

17. Analogical aids of deductive reasoning

The partial analogy between deduction and interrogation brings to the open another reason why the role of deductive logic in reasoning may appear smaller than it is. This time the reason is quite subtle. I have suggested elsewhere that the actual formal logic of quantification is not a very natural medium of reasoning for the human mind.[15] For it is a characteristic of the peculiarly human mode of information processing that it tries to avoid such nesting of quantifiers as may take place in the formal logic of quantification. For reasons which I shall not discuss here and which are related to von Neumann's observations concerning the differences between a computer and a brain, the human mind finds it difficult to operate with nested quantifiers.[16]

Such nesting of quantifiers is of course closely related to the increase in the number of individuals considered together which makes logical arguments theorematic.

So how can the human mind cope with highly theorematic modes of deductive reasoning? How has it handled best-known instances of quantificational reasoning in history, the arguments found in elementary geometry? I have suggested an answer elsewhere. What the human mind does is to enlist the help of analogical representations. For instance, in elementary geometry we use figures to illustrate the theorem to be proved and the argument that actually provides the proof. As philosophers of mathematics have emphasized in the last hundred-odd years, the use of such illustrations makes no difference to the validity of a geometrical argument, which can (and perhaps should) be carried out in purely formal terms.[17] But this correct observation closes the issue only if we are thinking in terms of definitory rules. For from the fact that figures or other analogical aids are dispensable from the definitory standpoint (i.e., from the standpoint of the validity of one's arguments), it does not follow that they are dispensable from the vantage point of strategic rules, at least as far as the ability of the human mind to discover and to master the arguments in question is concerned.

To put the point in other words, even though geometrical intuition is not needed to validate geometrical arguments, it may be needed to make it possible for the human reasoner to master them, even though they can in the last analysis be construed as being purely deductive. In general, the human mind frequently needs the help of analogical representations in order to handle even purely deductive arguments whose validity in no way depends on the analogical representations.

Now in applied deductive reasoning, such as is involved in my interrogative processes, the actual individuals obtainable via answers to wh-questions may provide such analogical representations as are often needed to facilitate the inquirer's deductive reasoning, assuming that the inquirer is a human being and not a computer. In other words, where deductive and interrogative inferences are parallel, interrogative ones are psychologically easier for us humans. The parallelism between deductive and interrogative steps helps to guide the search for the analogical representations which may be needed even when a conclusive argument could be carried out in strictly deductive terms. It is interesting to observe that here interrogative principles can serve as a guide to deductive strategies. The parallelism between the two thus supports traffic in both directions. What this amounts to is a reason why the role of deduction in inquiry can easily seem smaller than it is. For how can deductive inferences play a major role in actual substantial reasoning if they themselves often need the help of representations of the kind that can be provided by the rest of the inquiry, it may be asked. Natural though this question may be, it does not prove anything about the strategic role of logic in inquiry.

In historical terms, we have here an explanation of a striking fact. What for us are purely logical (quantificational) inferences have in certain cases been consistently called mathematical rather than logical modes of reasoning. This is especially true of those modes of reasoning which turn on existential instantiation and which were often given the label *ekthesis*.[18] They are precisely the kinds of inference for which the human mind typically needs analogical aids. Because of this need, they have not been considered purely formal, even though their validity is of course unaffected by the psychological need of "visual aids," "intuitions," or other analogical helpmates.

Historically speaking, we see here one of the most important causes of the underestimation of the role of logic in inference in general.

18. Critical thinking and deductive logic

My main result concerning the partial identity of deductive and interrogative strategies nevertheless needs a qualification or perhaps an amplification. The partial parallelism extends only as far as we can, or, rather, as far as the inquirer can, assume that nature's answers are unproblematically true. This assumption covers only a part of the uses of the interrogative model. Now many of my model's most novel and potentially most important applications are to situations in which the oracle's answers are not necessarily true, but have to be evaluated on the basis of the oracle's other answers plus, possibly, *a priori* knowledge concerning their reliability, e.g., knowledge of the probabilities of their being true.[19] Of course, the inquirer may need, and may also have, some information concerning the interdependencies of the different answers.

The interrogative model is obviously a most useful tool for the purpose of studying such kinds of inquiry, even though they have not yet been examined systematically from the vantage point of the model. Once again, questions of strategy will loom large. However, there no longer are any obvious analogies between the choice of such critical and evaluative strategies and the choice of moves in the "game" of formal logic (i.e., deductive strategy selection).

This amounts to a limitation of the role of formal logic in the general theory of reasoning. It therefore seems possible to conclude that logic in its customary forms plays a somewhat smaller role in critical and evaluative argumentation than in constructive reasoning. Or, I should perhaps say, there exists a sharper division of labor between the logical and the empirical aspects of inquiry on the critical than on the constructive side of the overall enterprise.

Jaakko Hintikka

NOTES

1. For Newton, see Jaakko Hintikka and James Garrison, "Newton's Methodology and the Interrogative Logic of Experimental Inquiry," forthcoming in the proceedings of the Spring 1987 workshop on "300 Years of the *Principia*—Realism Then and Now," ed. by Zev Bechler *et al.*

2. See Jaakko Hintikka, *Logic, Language-Games and Information*, (Oxford: Clarendon Press, 1973). I have sometimes also used the terms "surface tautology" and "depth tautology" for the two parties of the distinction.

3. See Jaakko Hintikka, "C. S. Peirce's 'First real Discovery' and its Contemporary Significance," in *The Relevance of Charles Peirce*, (La Salle, IL: The Hegeler Institute, 1983), 107–18.

4. For if there were a recursive function which would give an upper bound to the number of new individuals (existential instantiations) needed in the proof, we easily could construct a (finite) upper bound to the length of the prospective proofs which would lead from T to C. By constructing all the potential proofs of this length, we could decide effectively whether C follows logically from T. But such a decision method is known to be impossible.

5. See E. W. Beth, "Semantic Entailment and Formal Derivability," *Mededelingen van de Koninklijke Nederlandse Akademie van Wetenschappen*, Afd. Letterkunde, N.R. vol. 18, no. 13, (1955).

6. But who among actual scientists ever carried out his or her experimental arguments as if they were logical proofs? An answer is easy: Isaac Newton, for one. His *Optics* is constructed completely *à la* Euclid, with axioms, postulates, definitions, propositions, theorems, problems, and proofs. The only main difference as compared with Euclid is that Newton occasionally inserts into his exposition what he calls "a proof by experiment." In other words, he describes an experiment and adds its result as a fresh premise. This is *mutatis mutandis* precisely what the interrogative model suggests.

7. See, e.g., Jaakko Hintikka "What Is the Logic of Experimental Inquiry?", *Synthese* vol. 74, no. 1 (1988): 173–90.

8. For a fine exposition of the "arbitrary objects" idea, see Kit Fine, *Reasoning With Arbitrary Objects* (Oxford: Basil Blackwell, 1985). For some of the philosophical problems associated with the idea, see *op. cit.*, n2 above, pp. 109–14.

9. See R. G. Collingwood, *Essay on Metaphysics* (Oxford: Clarendon Press, 1940); Hans-Georg Gadamer, *Truth and Method* (New York: Continuum, 1975), especially pp. 333–41.

10. See here Jaakko Hintikka, "The Fallacy of Fallacies," *Argumentation* vol. 1 (1987), 211–38.

11. Cf. here Jaakko Hintikka, "Aristotle's Incontinent Logician," *Ajatus* 37 (1978), 48–65.

12. Cf. *op. cit.*, n2 above, 201–05.

13. *The Critique of Pure Reason*, second ed., xii–xiii.

14. See M. E. Szabo, *The Collected Papers of Gerhard Gentzen* (Amsterdam: North-Holland, 1969).

15. In "A Note on Anaphoric Pronouns and Information Processing by Humans," *Linguistic Inquiry* 18 (1987), 111–19.

16. See John von Neumann, *The Computer and the Brain* (New Haven, CT: Yale University Press, 1958). The connection with my ideas is this: von Neumann considers the nesting of functions as the crucial obstacle to information-processing by humans. Now when existential quantifiers are replaced by what are known as Skolem functions, the kind of quantificational complexity I have in mind here simply becomes an instance of the von Neumann-type nesting of functions.

17. Cf. *op. cit.*, n2 above, pp. 208–11.

18. See *op. cit.*, n2 above, pp. 213–18.

19. This extension of the interrogative model is introduced and briefly discussed in Jaakko Hintikka, "The Interrogative Approach to Inquiry and Probabilistic Inference," *Erkenntnis* 26 (1987), 429–42.

3

* INTERROGATIVE LOGIC AS A GENERAL THEORY OF REASONING

1. THE GENERALITY OF INTERROGATIVE REASONING

The interrogative approach to reasoning and argumentation is not just one approach to its subject among many. Both historically and systematically, it is arguably the first and foremost theory of reasoning.

Historically, the first model of reasoning to capture philosophers' attention was the Socratic method of questioning or *elenchus*. Plato systematized it into a practice of questioning games used in the Academy to train his students in philosophical and scientific reasoning. Aristotle developed in the *Topica* a theory of such interrogative argumentation, emphasizing among other things its usefulness for scientific purposes. In so doing, his attention was caught by answers that are necessitated by earlier ones or, as we would express it, are logically implied by them. His study of such answers became the first study of logic in history. (Cf. here Hintikka 1996(a).)

Thus the interrogative method is historically speaking not just one way of approaching logic and argumentation. It is *the* theory of reasoning, logical as well as empirical, comprising deductive logic as a special case.

Systematically, too, the interrogative approach can be argued to be a general theory of reasoning. In order to see the sense in which this is the case, consider someone's line of reasoning, expressed in the form of a series of propositions (sentences). What is a new proposition in such a series like? Either the information it codifies is contained in the earlier propositions in the series, or it is not. In the former case, the step is a logical (deductive) consequence of these earlier steps. Such argumentative steps are to be studied by means of deductive logic.

In the latter case, if the arguer is rational, he or she must know where the new information comes from. Otherwise we are not dealing with reasoning and inference but mere guesswork. We will use as a technical term for all sources of information the word *oracle*. Since the information is new, the inquirer must somehow have received this information as a response to his or her own initiative, which is an action directed to some particular oracle. Since this source of information is therefore known, we might as well think of the new information as an answer to a question the inquirer has addressed to the oracle in question. But what this means is that such a rational line of reasoning can in principle be thought of as an interrogative process. The Greeks were right: all reasoning can be thought of as questioning. The main superficial change as compared with Socratic

*Written jointly with Ilpo Halonen and Arto Mutanen.

elenchus is that purely logical (deductive) steps are in our treatment distinguished from interrogative ones. This modification was anticipated by Aristotle, even though it is not realized clearly enough by most historians that he still considered even syllogistic inferences in science merely as special kinds of question-answer steps in interrogative inquiry.

2. THE BASIC RULES OF INTERROGATIVE LOGIC

What, then, is the logic (structure) of interrogative inquiry? We will first consider a simple case which nevertheless brings out several important features of reasoning and inquiry in general. This case is characterized by the following features:

(i) There is only one oracle.
(ii) The set of answers the oracle will provide remains constant throughout the inquiry.
(iii) All of the oracle's answers are true, and known by the inquirer to be true.

Given these assumptions, the logic of interrogative reasoning will at first sight seem exceedingly simple. We can formulate the rules for such a procedure in the form of rules for a game played by the inquirer against an oracle. The book-keeping method we use is a variant of Beth's (1955) *tableau* method in deductive logic. All the moves are initiated by the inquirer. The only role that the oracle plays is to respond to the inquirer's questions. Initially, there are certain initial premises on the left side of the *tableau* and the proposition to be interrogatively established (proved) on the right side. There are two kinds of moves, logical inference moves and interrogative moves. The logical inference moves are simply a variant of the *tableau*-building rules of the usual *tableau* method. For simplicity, we will here and in the rest of this paper assume that all the formulas we are dealing with are in the negation normal form, i.e. that all negation signs precede immediately atomic formulas or identities, unless otherwise indicated. The basic concepts of the *tableau* method, such as column, closure and *subtableau*, are used as usual. An example of the *tableau*-building rules is the following:

If $(S_1 \vee S_2)$ occurs on the left side of a *subtableau*, it is divided into two *subtableaux* which have as a new formula in their left column S_1 and S_2, respectively.

In order to obtain a more concise notation we will formulate the *tableau*-building rules as inverses of the corresponding Gentzen-style (sequent) rules. The formulas on the left side of a *(sub)tableau* go to the antecedent of the corresponding sequent, the formulas on the right side go to the consequent. Thus in this notation the disjunction rule becomes

(L.∨) $\dfrac{\Gamma,(S_1 \vee S_2) \rightarrow \Delta}{\Gamma,S_1 \rightarrow \Delta \quad \Gamma,S_2 \rightarrow \Delta}$.

The other rules can be formulated as follows:

(L.&) $\dfrac{\Gamma,(S_1 \,\&\, S_2) \rightarrow \Delta}{\Gamma,S_1,S_2 \rightarrow \Delta}$,

(L.e) $\dfrac{\Gamma,(\exists x)S[x] \rightarrow \Delta}{\Gamma,S[b] \rightarrow \Delta}$

where b is a new individual constant. Constants introduced by this rule are called *dummy names*.

(L.a) $\dfrac{\Gamma,(\forall x)S[x] \rightarrow \Delta}{\Gamma,(\forall x)S[x],S[b] \rightarrow \Delta}$

where b is an individual constant occurring in Γ, Δ or S[x].

In fact, (L.a) is a special case of a more general rule:

(L.A) $\dfrac{\Gamma,(\forall x)S[x] \rightarrow \Delta}{\Gamma,(\forall x)S[x],S[t] \rightarrow \Delta}$

where t is a term built of function constants and individual constants occurring in Γ, Δ or S[x].

The right-hand rules are mirror images of (L.∨)-(L.A), e.g.,

(R.∨) $\dfrac{\Gamma \rightarrow \Delta,(S_1 \vee S_2)}{\Gamma \rightarrow \Delta,S_1, S_2}$.

For formulas which are not in a negation normal form, we will use rules to drive negation deeper into the formulas in question, e.g.,

(L.~∨) $\dfrac{\Gamma,\sim(S_1 \vee S_2) \rightarrow \Delta}{\Gamma,(\sim S_1 \,\&\, \sim S_2) \rightarrow \Delta}$.

We also need some suitable structural rules, which allow us to permute the members on the left side or on the right side of a sequent, and closure rules. The following can serve as rules for closure:

(C.T) A tableau is closed iff all its *subtableaux* are closed.

A *subtableaux* is closed if it includes one of the following:

(C.L) $\Gamma, S_1, \sim S_1 \rightarrow \Delta$
(C.R) $\Gamma \rightarrow \Delta, S_1, \sim S_1$
(C.LR) $\Gamma, S_1 \rightarrow \Delta, S_1$

If identity is included in the underlying language, we need substitution rules. These rules are illustrated by the following examples:

(LL.=) $\dfrac{\Gamma, S[a], (a=b) \rightarrow \Delta}{\Gamma, S[a], S[b], (a=b) \rightarrow \Delta}$,

(LR.=) $\dfrac{\Gamma, (a=b) \rightarrow \Delta, S[a]}{\Gamma, (a=b) \rightarrow \Delta, S[a], S[b]}$.

We also need closure rules for identity, for instance a rule that says that a *subtableaux* is closed if it contains

(C.L=) $\Gamma, \sim(a=a) \rightarrow \Delta$

For reasons that will become clear later, it is advisable to generalize (L.e) and (R.a) as to allow functional instantiation over and above the usual existential instantiation:

(L.E) $\dfrac{\Gamma, S_0[(\exists x)S_1[x]] \rightarrow \Delta}{\Gamma, S_0[S_1[f(y_1,\ldots,y_n)]] \rightarrow \Delta}$

where f is a new function constant and $(\forall y_1),\ldots,(\forall y_n)$ are all the universal quantifiers within the scope of which $(\exists x)S_1[x]$ occurs in S_0. Function constants introduced in two way are said to represent *dummy functions*.

(R.A) Analogously.

It is important to recall here and in similar rules below that S_0 is assumed to be in the negation normal form, i.e. that all negation signs in S_0 are in front of atomic formulas. (L.e) is taken to be a special case of (L.E).

Likewise it is advisable to generalize (L.$\vee$) and (R.&) an analogous way:

(L.∨f)
$$\frac{\Gamma, S_0[(S_1 \vee S_2)] \rightarrow \Delta}{\Gamma, S_0[(S_1 \,\&\, f(y_1,\ldots,y_n)=0) \vee (S_2 \,\&\, f(y_1,\ldots,y_n)\neq 0)] \rightarrow \Delta}$$

Here f is a new function constant (dummy function) and $(\forall y_1),\ldots,(\forall y_n)$ are all the universal quantifiers within the scope of which $(S_1 \vee S_2)$ occurs in S_0.

(R.&f) Analogously.

The old rules (L.∨) and (R.&) are taken to be special cases of the new rules (L.∨f) and (R.&f).

3. RULES FOR QUESTIONING

The rule for questioning can be formulated as follows:

(L.Q) If the presupposition of a question occurs on the left side of a *subtableau*, the inquirer may address the corresponding question to the oracle. If the oracle answers, the answer is added to the left side of the *subtableau*.

For obvious reasons, there is no right-side questioning rule. Equally obviously, no dummy names must occur in an answer.

The rule (L.Q) refers to the notion of presupposition. Hence our next task is to explain this notion. We will at the same time define (tentatively) another basic notion from the theory and practice of questions, the notion of answer. In so doing, we have to distinguish different kinds of questions from each other. In a propositional question, the *presupposition* is a disjunction

$(S_1 \vee \ldots \vee S_n)$

and the possible *answers* are $S_1,\ldots,S_n$. In a (nonmultiple) wh-question, what the questioner is looking for is an individual, possibly depending on others. The *presupposition* of such a question (assuming it is in negation normal form) can be any sentence containing an existential quantifier or disjunction, say

(3.1) $S_0[(\exists x)S_1[x]]$.

An *answer* to the question is then of the form

(3.2) $S_0[S_1[f(y_1,y_2,\ldots,y_n)]]$

where $(\forall y_1),\ldots,(\forall y_n)$ are all the universal quantifiers within the scope of which $(\exists x)$ occurs in (3.1).

Of course, what in ordinary discourse is called an answer is often not a sentence like (3.2) but the substituting function f or a substituting term.

What this means is that we interpret wh-questions as "for instance" questions. The questioner is not trying to identify all the individuals satisfying a certain condition. Coming to know one such individual will satisfy the inquirer. Likewise, in the propositional question whose presupposition is $(S_1 \vee S_2 \vee \ldots \vee S_n)$ one of the alternatives $S_1, S_2, \ldots, S_n$ is enough to serve the questioner's aims. Questions in which the purpose is to specify all the true or true-making alternations can be dealt with by the same conceptual tools as we are using here, although we will not do so here.

It is to be noted that there are also questions where the alternatives are propositions, but propositions that depend on certain individuals. For such questions can be taken to be of the form

$$S_0[(S_1 \vee \ldots \vee S_n)],$$

for instance

$$S_0[(S_1 \vee S_2)]. \tag{3.3}$$

An answer (in our terminology) is then of the form

$$S_0[(S_1 \,\&\, f(y_1,\ldots,y_n)=0) \vee (S_2 \,\&\, f(y_1,\ldots,y_n)\neq 0)] \tag{3.4}$$

where $(\forall y_1),\ldots,(\forall y_n)$ are all the universal quantifiers within which $(S_1 \vee S_2)$ occurs in (3.3). When there are no such universal quantifiers, there are two possible answers, viz. $S_0[S_1]$ and $S_0[S_2]$.

The specification of the set Φ of the answers that an oracle will give is a part of the specification of a specific game of interrogative inquiry. It is assumed that Φ remains constant throughout an inquiry, and that an oracle will always answer a question when one of its answers is in Φ. It is also assumed that the new individual and function constant introduced by rules like (L.E) and (R.A) do not occur in the members of Φ.

Together these rules define certain interrogative games (types of interrogative inquiry). However, strictly speaking they define only a cut-free version of interrogative inquiry. A more general type of interrogative inquiry is then defined by adding to rules so far introduced the following ones:

(L.taut) $$\frac{\Gamma \rightarrow \Delta}{\Gamma,(S \vee \sim S) \rightarrow \Delta}$$

(R.cont) $\dfrac{\Gamma \to \Delta}{\Gamma \to \Delta,(S \,\&\, {\sim}S)}$.

Here S is an arbitrary formula of the underlying language. (L.taut) is to all practical purposes the same as proof theorists' "cut rule."

We will call an interrogative logic which includes (L.taut) and (R.cont) an *extended interrogative logic*.

All the rules for logical inferences (tableau building) and for questioning have been formulated here with an eye on a situation in which all answers are true and known to be true. We can call this case the logic of discovery. It turns out, however, that the same rules are basic for all interrogative logic. The main qualification needed in order to make this claim is that in a fully explicit interrogative logic we have to take into account also the epistemic element. (See sec. 12 below.)

The interrogative derivability of C from the initial premises T in a model M will be designated by

(3.5) M:T |– C.

This notion generalizes in an interesting way both the familiar notion of truth in a model, i.e. M |= C, and the equally familiar notion of logical consequence, T |– C. The former relation is obtained as a special case of (3.5) when T is empty and all questions are answerable. The latter is obtained when no questions can be asked. These observations throw some light on the idea of truth as the ideal limit of inquiry.

4. SOME MAIN RESULTS

These rules, as they are so far formulated, capture only the simplest case of interrogative inquiry, viz. the case in which a single oracle's answers are all true (and known to be true by the inquirer). This might be considered as the case of pure discovery. The fact that we can develop an interesting theory of such discovery procedures is already highly suggestive, for among other things it shows conclusively the possibility of a genuine logic of discovery.

Such a logic is made interesting among other things by the possibility of establishing interesting metatheorems about the logic of interrogative inquiry. Some of them are extensions of the crucial metatheoretical results from logical theory to the logic of interrogative reasoning. Some are technically trivial, but interesting because of their applications.

Completeness Theorem: If Φ is the set of all available answers and T the set of initial premises, then a conclusion C can be established in M by means of extended interrogative logic iff $(\Phi \cup T) \vdash C$.

Proof: Let Φ be the set of all answers of the oracle.
Let us suppose that

$$M{:}T \vdash C.$$

In interrogative derivation of C from T in M we use logical moves and interrogative moves. If there is no interrogative move then there is nothing to prove. Let $S_1,\ldots,S_n$ be the answers that we use in an interrogative derivation of C from T in M. So we have

$$T \cup \{S_1,\ldots,S_n\} \vdash C.$$

But by the definition of the set Φ, $S_1,\ldots,S_n \,\varepsilon\, \Phi$. Hence we have

$$T \cup \Phi \vdash C.$$

This proves the "only if" part. Let us suppose conversely that

$$\Phi \cup T \vdash C.$$

By compactness of first-order logic there is a finite set $\Delta \subseteq \Phi$ such that

$$\Delta \cup T \vdash C.$$

Let $\Delta = \{S_1,\ldots,S_n\}$. Now the only problem is how we can have the sentence S_i as an answer to a question addressed to the oracle (for $i = 1,2,\ldots,n$). Sentences $S_1,\ldots,S_n$ do not contain any dummy names that are introduced by the rules (L.Ef) or (R.Af). By using the rule (L.taut) we can introduce the tautologies $(S_i \vee {\sim}S_i)$ $(i=1,2,\ldots,n)$ to the left side of subtableau. So the inquirer can ask the questions 'S_i or ${\sim}S_i$?' $(i=1,2,\ldots,n)$. Now since $S_i \varepsilon \Delta \subseteq \Phi$ and all the sentences in the set Φ are true in the underlying model, we have ${\sim}S_i\Phi$. Hence we have the sentence S_i as an answer to the corresponding question $(i=1,2,\ldots,n)$. Hence

$$M{:}T \vdash C.\square$$

The question of the completeness or incompleteness of unextended interrogative logic is a much messier one. It will not be discussed here.

Other results include the following:

Yes-No Theorem: In the extended interrogative logic, if M:T |– C, then the same conclusion C can be established by using only yes-no questions.

A terminological explanation is in order here. For the propositional question 'Is it the case that S_1 or...or S_n?' the presupposition is ($S_1 \vee ... \vee S_n$). We say that a propositional question whose presupposition is of the form ($S \vee \sim S$) is a *yes-no question.*

Proof: Let us suppose that the answer to a propositional question 'S_1 or ... or S_n?' is the sentence S_i ($i \leq n$). Now instead of asking the propositional question, the inquirer can introduce the tautology ($S_i \vee \sim S_i$) by using the rule (L.taut) and use it as the presupposition of the yes-no question 'S_i or $\sim S_i$?'. Clearly the sentence S_i is an available answer to this question, for it was answerable as an answer to the given propositional question.

Let us suppose that the answer to the wh-question '$(\exists x)S[x]$?' is the sentence S[b]. Now instead of asking the wh-question, the inquirer can introduce the tautology ($S[b] \vee \sim S[b]$) by using the rule (L.taut) and use it as the presupposition of the yes-no question 'S[b] or ~S[b]?'. As in the case of any propositional question, the sentence S[b] is available as an answer to that question.

The other cases can be dealt with similarly.□

Covering Law Theorem (special case): Let T be a theory in the first-order language L and P be a one-place predicate symbol occurring in T. Let M:T |– P(b) where b is a given individual constant be established by means of a set of answers A. Assume that A is consistent and that T does not entail P(b). If P does not occur in A and if b does not occur in T, then there is a formula H[x] such that

(a) $T \vdash (\forall x)(H[x] \supset P(x))$;
(b) $A \vdash H[b]$;
(c) Neither P nor b occurs in H[x].

Proof: Let us suppose that M:T |– P(b). Then obviously

$A \cup T \vdash P(b)$.

By the compactness of first-order logic there is a finite subset T* of T such that

$A \cup T^* \vdash P(b)$.

Let $\wedge T^*$ be the conjunction of the members of the set T*. Then we have

(4.1) $A \cup \{\wedge T^*\} \vdash P(b)$.

Now the deduction theorem of first-order logic gives us

(4.2) $A \vdash (\wedge T^* \supset P(b))$.

Here Craig's (1957 (a) and (b)) interpolation theorem can be seen to apply. Hence there exists an interpolant I[b] such that

(4.3) $A \vdash I[b]$ and

(4.4) $I[b] \vdash (\wedge T^* \supset P(b))$.

By the principles of first-order logic we obtain from (4.4)

$\wedge T^* \vdash (I[b] \supset P(b))$.

Since b does not occur in T*, we have

(4.5) $\wedge T^* \vdash (\forall x)(I[x] \supset P(x))$.

Hence we have

(4.6) $T \vdash (\forall x)(I[x] \supset P(x))$.

The vocabulary of I[b] is shared by A and $(\wedge T^* \supset P(b))$. Thus P does not occur in I[b] since it does not occur in A. Trivially b does not occur in I[x]. We get the intended result by putting I[x] = H[x].□

Covering Law Theorem (suitably generalized) has important application in the philosophy of science, in particular in the theory of explanation. It can also be used to elucidate the nature of why-questions and of their role in interrogative inquiry. (See sec. 19 below and Hintikka and Halonen 1995.)

Covering Law Theorem (general version): Let T be a theory in the first-order language L and $F = F[b_1,...,b_n]$ be a sentence such that $F[x_1,...,x_n]$ is a formula of L and $b_1,...,b_n \in$ do(M). Let $M:T \vdash F$ be established by means of a constant set of answers A. Let us assume that A is consistent and that T does not entail $F[b_1,...,b_n]$. If other constants than $b_1,...,b_n$ of F do not occur in A and if $b_1,...,b_n$ do not occur in T, then there is a formula $H[x_1,...,x_n]$ of L such that

(a) $T \vdash (\forall x_1)...(\forall x_n)(H[x_1,...,x_n] \supset F[x_1,...,x_n])$
(b) $A \vdash H[a_1,...,a_n]$
(c) none of the constants of $F[x_1,...,x_n]$ occur in $H[x_1,...,x_n]$.

Proof: By assumption

$A \cup T \vdash F[b_1,...,b_n]$.

By the compactness of first-order logic there is a finite subset T* of T such that

$A \cup T^* \vdash F[b_1,...,b_n]$.

In the same way as we obtained the formula (4.1) in a special case we obtain here

(4.7) $A \vdash (\wedge T^* \supset F[b_1,...,b_n])$

where $\wedge T^*$ is as in (4.1). By using Craig's interpolation theorem (1957(a) and (b)) there exists an interpolant $I[b_1,...,b_n]$ such that

(4.8) $A \vdash I[b_1,...,b_n]$ and
(4.9) $I[b_1,...,b_n] \vdash (\wedge T^* \supset F[b_1,...,b_n])$.

Just as we obtained (4.6) we can here obtain

(4.10) $T \vdash (\forall x_1)...(\forall x_n)(I[x_1,...,x_n] \supset F[x_1,...,x_n])$.

The vocabulary of $I[b_1,...,b_n]$ is shared by A and $(\wedge T^* \supset F[b_1,...,b_n])$. Thus other constants than $b_1,...,b_n$ of $F[b_1,...,b_n]$ do not occur in $I[b_1,...,b_n]$ since they do not occur in A. Constants $b_1,...,b_n$ do not occur in $I[x_1,...,x_n]$. We obtain the intended result by putting $H[x_1,...,x_n] = I[x_1,...,x_n]$.□

The last result presented here will be called the *Strategy Theorem*. It is formulated only informally. (Cf. here Hintikka 1989.) It requires a preliminary explanation. If you review the questioning rule (L.Q), you can see that the formulas that can serve as presuppositions of questions are precisely the same formulas as the ones to which (L.E) or (L.∨f) can be applied. Hence there is a parallelism between the choice of one's interrogative strategy and the choice of one's deductive strategy. The question as to which formula to use as the presupposition of one's next question is paralleled by the question as to which formula to apply (L.E) or (L.∨f) to next. The Strategy Theorem says that the two choices are governed by the same principles.

Above we have stated the rules which define the moves that may be carried out in deductive reasoning and in interrogative reasoning. These rules tell only how to reason (deductively and/or interrogatively) correctly. We will say that these rules are *definitory rules* for deduction and interrogation, respectively. These rules tell what is permitted in a game of reasoning. But we need also different kinds of rules which define how to reason well, i.e. we need rules which tell how to use definitory rules effectively. We will call such rules *strategic rules* for deduction and interrogation, respectively. Such rules tell a reasoner how to argue effectively. Unfortunately there are no recursive rules which can be applied in every case. But it is possible to formulate some general strategic principles in deduction and interrogation. Among other things, we can establish a certain strategic parallelism between deduction and interrogation. (For an earlier discussion of the same idea, see Hintikka 1989.)

At each stage of an interrogative argument, only a finite number of applications of the rule (L.E) are possible. The sentence on the left side of the sequent reached at the time to which (L.E) can be applied are precisely the same ones as can serve as presuppositions of wh-questions. From studies of strategies of theorem-proving it is known that the crucial strategic question in deduction is which of these sentences to apply (L.E) to first. (Thus we can disregard the choice of propositional questions.) Obviously, in interrogative reasoning the crucial question is which of them to use first as the presupposition of a wh-question. With certain qualifications, it can be seen that not only is the range of choice the same in (L.E) and wh-questions. What is crucial, the optimal choice is the same in both cases.

This can be seen by examining the parallelism between deductive and interrogative steps. Suppose, for instance, that there occurs on the left side of a *(sub)tableau* an existential sentence $(\exists x)S[x]$. In deduction the rule (L.e) says that the inquirer may add the sentence $S[\alpha]$ to the left side of a *(sub)tableau* where α is a new dummy name.

In interrogation the inquirer may use the rule (L.Q) and ask the wh-question '$(\exists x)S[x]$?'. If the oracle answers then the inquirer may add the sentence S[b] to the left column of the (sub)tableau where b is some individual from the domain of the underlying model.

Consider the difference between deduction and interrogation in the treatment of existential sentences. First, everything that can be done with $S[\alpha]$ can be done with S[b], but not vice versa. The reason is that b is a real individual and α is only a dummy name. This means that we can ask questions concerning the individual b but not concerning the dummy name α. To see what this means, let us suppose that there occurs on the left side of the same (sub)tableau another existential sentence $(\exists x)S_1[x]$. In deduction the inquirer uses again the rule (L.e) and adds a new sentence $S_1[\beta]$ to the left side of the (sub)tableau where β is a new dummy name. In interrogation the inquirer may use information that he already has and use the rule (L.Q) to ask the yes-no question '$S_1[b]$ or $\sim S_1[b]$?'. But in a similar way as above we see that if we use the rule (L.e) in deduction, then in interrogation we use the rule (Q.L) (if the question is answerable), and vice versa.

The general case $S_0[(\exists x)S[x]]$ can be treated similarly.

We can see that there are some differences between strategies in deduction and in interrogation. But we see that always when the inquirer in deduction uses in an optimal strategy one of the rules (L.E) or (L.$\vee$f), then the inquirer in interrogation should use the rule (L.Q) and ask the corresponding wh-question or propositional question, and vice versa.

Thus we have made plausible the following metatheoretical result:

Strategy Theorem: The optimal choice of the wh-question to ask is the same as the optimal choice of the rule (L.E) to apply, assuming that the question is answerable.

In this form, the Strategy Theorem applies in the first place to unextended interrogative logic. It can be extended further by noting that (L.taut) and (R.cont) can be used in ordinary first-order logic to shorten proofs. In interrogative logic, they open the possibility of raising new yes-no questions and thereby facilitating new lines of reasoning. How closely parallel the principles of choosing the tautology for the two purposes are remains to be studied.

The Strategy Theorem illustrates one of the major directions into which we are pushed in trying to develop a theory of interrogative reasoning. In doing so, we are soon led to realize the importance of strategies and strategic thinking, not only in the applications of interrogative logic, but in the very theory of this logic.

The strategic viewpoint also helps us to put into the proper perspective the Yes-No Theorem. It is virtually trivial technically and yet it seems to have sweeping implications. What are other kinds of questions good for, anyway, if their job could equally well be done by the simplest of all questions, yes-no questions?

The answer is that the purely inferential job of other kinds of questions can be done by yes-no questions, but that their strategic job cannot be so done. As one can see from the proof of the Yes-No Theorem, the yes-no question which does the same job as another one, say a simple wh-question, is found by means of the answer, say S[b], to the wh-question. Clearly by asking "Is S[b] the case or not?" the inquirer can obtain the same statement as an answer to the yes-no question. But how is the interrogative reasoner supposed to know to ask this particular question and not, say, a question about S[d]? Typically one does *not* know it without the help of an earlier answer. Hence the "can" in the Yes-No Theorem means a definitory "can," not a strategic "can." It's import can perhaps be best expressed by saying that any interrogative argument can in the light of hindsight be subsequently replaced (in extended interrogative logic) by an argument where all questions used are yes-no questions.

5. A HIERARCHY OF LOGICS OF DISCOVERY

Even though our pure logic of discovery might at first sight appear exceedingly simple, it soon turns out to be a rich and intricate logical theory. The mere existence of such a logical theory amply shows that a logic of discovery in a perfectly natural sense is possible, even though the impossibility of such a logic has for a long time been a dominant dogma in the philosophy of science. Indeed, it can be suggested that the logic of epistemological justification is more complicated and more difficult to have an overview on than the pure logic of discovery.

There is nevertheless no such thing as *the* logic of discovery. For one important thing, the structure of an interrogative process depends crucially on the class of answers Φ which the oracle(s) will give. (It is assumed to be constant throughout an interrogative inquiry.) How is Φ determined? In applications, there are usually various *ad hoc* restrictions on Φ which are hard to reach any overview on. For instance, in the case of scientific reasoning their study involves questions of observational and experimental methodology.

What can be studied here by means of logical methods is how our interrogative logic depends on the structural restrictions on available answers. The effects of such restrictions can be seen by examining the limiting case in which the structural restrictions are the only relevant ones. In this direction, the most important dimension is the quantificational complexity of the answers one's oracle (or oracles) can provide. (See Hintikka 1985.) The simplest case is the one in which all available answers are quantifier-free, i.e., particular propositions. In view of the Yes-No Theorem this case is to all practical purposes the one in which we restrict our attention to questions of the form "A or not-A?" where A is an atomic proposition. The assumption that we can restrict our attention in this way may be called the Atomistic Postulate. However, less narrow assumptions are likewise possible. The next case may be taken to be the one in which the available answers are either quantifier-free or of the form

$$(\forall x_1)(\forall x_2)...S[x_1,x_2,...,a_1,a_2,...]$$

where S is quantifier-free. We will call this the A-case. The next one is the AE-case in which answers are also allowed to be of the form

$$(\forall x_1)(\forall x_2)...(\exists y_1)(\exists y_2)...S[x_1,x_2,...,y_1,y_2,...,a_1,a_2,...].$$

Sentences of this form are said to be AE sentences.

This hierarchy can obviously be continued. It is of a great importance for the logic of interrogative inquiry. Among other things, it enables us to characterize different types of inquiry from each other and likewise to classify different oracles according to the maximal logical complexity of the answers that they will give. For instance, purely observational inquiry is characterized by the Atomistic Postulate. In a diagnostic inquiry in medicine one oracle (the patient) provides only particular answers, whereas others (medical handbooks, the doctor's own background, biological knowledge, etc.) can in principle provide answers of any logical complexity.

6. THE ATOMISTIC CASE

The atomistic case results when the Atomistic Postulate is assumed. The result is a clear-cut and theoretically manageable logic of discovery. Let us assume that our interrogative logic is applied to a model ("world") M, and let us follow the usual practice of calling the set D(M) of negated or unnegated atomic sentences true in M the *diagram* of M. (It is assumed that each member of the domain of individuals do(M) of M has a name in the underlying language.) In the old-fashioned terminology employed by Carnap, D(M) is essentially the state-description true in M. If no other restrictions are imposed on answers, we have M:T |– C if and only if {D(M) ∪ T} |– C.

In the atomistic case, the theory of interrogative logic thus reduces largely to ordinary logic, including its model theory. This enables us to establish a number of clear-cut results about the atomistic case. As an example, we can consider the question as to what an optimal theory looks like in interrogative logic. (Cf. here Hintikka 1984.) On the basis of ordinary deductive logic, an ideal theory T is of course a deductively complete theory, that is, a theory so that T |– C or T |– ~C for every closed sentence C of the underlying language. Accordingly to what was just said, in interrogative logic the optimal theory T is such that

{D(M) ∪ T} |– C or
{D(M) ∪ T} |– ~C

for each C in the underlying language (including names for the members of the domain do(M) of M) and for each model M of T. A moment's thought shows that such a theory is precisely what in model theory is known as a model-complete theory. (Cf. Robinson 1963.) Hence model-complete theories are the optimal ones if the Atomistic Postulate is applicable.

Most epistemologists and philosophers of science have tacitly assumed that the Atomistic Postulate is applicable in empirical reasoning. In such a case, the relevant oracle is nature. Admittedly, it looks plausible to assume that nature only tells us what happens on some particular occasion or other. She never gives general answers; she does not for instance tell what happens always and everywhere.

The Atomistic Postulate nevertheless soon leads into trouble or perhaps rather into a problem. For how can we then account in empirical terms for the kind of reasoning that leads to general theories? From particular propositions alone one cannot logically derive general theories. For the purpose, we also need strong initial premises T. But where do they come from? They cannot any longer be obtained empirically, for they are needed for the process of theory formulation. Hence what is needed is apparently some nondeductive principles of reasoning. These principles are usually referred to as principles of inductive inference. Thus one possible reaction to the problems caused by the adoption of the Atomistic Postulate is an inductivistic epistemology.

An inductivistic epistemology falls outside the scope of our discussion, however. Or, rather, the interrogative logic developed here is calculated among other things to replace inductivistic methodology, not to embody it. (Cf. below sections 8 and 17.) We are not adopting the most popular alternative to inductivistic model of reasoning in epistemology, either, which is the hypothetic-deductive model of theoretical reasoning. According to this idea, the formation of general theories is not a rational process that can be dealt with any kind of "logic of discovery." It depends on intuition and serendipity rather than logic. Its results can be tested by comparing the deductive consequences of the resulting theory with experience, but the original discovery cannot be dealt with by means of a rational theory. In our view, such a view amounts to a failure to deal with one of the most important questions concerning human reasoning.

7. ON THE LOGIC OF EXPERIMENTAL REASONING

Our interrogative logic shows us a way out of this entire problem situation. The culprit here is not the problem of induction or the impossibility of the logic of discovery. It is the Atomistic Postulate. It simply is not true that the answers an inquirer receives from nature in an empirical inquiry are all particular propositions. The simplest counter-example is offered by controlled experiments. (They have in fact been considered ever since Kant as typical examples of nature's answers to an inquirer's questions to nature; see *Critique of Pure Reason* B xiii-xiv.) In a controlled experiment, a scientist is asking how the observed variable y changes when the controlled variable x is varied. An answer to such a question is an expression of functional dependence which can be taken to have the logical form of a sentence of the AE form, e.g. of

$$(\forall x_1)(\forall x_2)...(\exists y_1)(\exists y_2)...S[x_1,x_2,...,y_1,y_2,...,a_1,a_2,...].$$

Hence the logic of experimental inquiry is an AE-logic rather than atomistic logic. It may even be argued that more complex propositions can be obtained as answers to experimental questions to nature. (Cf. here Hintikka 1988.)

The true logic of discovery therefore has at least the AE complexity. This does not mean that it cannot be studied in the same way as the logic of atomistic inquiry. The difference lies in the fact that instead of the diagram D(M) of the relevant model we have to consider the set $D^{AE}(M)$ of all sentences of the AE form true in M — or perhaps some set which includes even more complex sentences.

There are results in effect concerning the interrogative AE logic in the literature, but it is not very easy to obtain a good overview on it. In any case, the AE logic is one of the most interesting and important forms which the logic of discovery can take. Its Protean character is an indication of its having been able to capture something of the richness of real-life nontrivial reasoning.

8. INTERROGATIVE INQUIRY BY MEANS OF UNCERTAIN ANSWERS

The basic interrogative logic so far developed is nevertheless too simple in several different ways. First, we must give up the assumption that all of the oracle's answers are true. This can be done surprisingly simply by rules that merely allow the inquirer to "bracket" any earlier answer (or initial premise) and conversely to "unbracket" a previously bracketed answer in the game *tableau*. The only catch here is that when an answer (or initial premise) is bracketed, all subsequent steps which depend on it must also be bracketed. This motivates the following rules which supplement the earlier ones:

(Q1) *Rule for Bracketing Lines:* At any stage of an interrogative game the inquirer may bracket any earlier answer or initial premise. We might for instance use a special kind of bracket which will normally signal that the inquirer thinks the answer or premise is uncertain.

(Q2) *Rule for Bracketing Lines Derived from Bracketed Lines:* If a line in the table of an interrogative game is bracketed, all later lines dependent on it must also be bracketed.

(Q3) *Rule for Not Using Bracketed Lines in the Table:* When a line is bracketed, it may not be used in any application of the game rules, including the closure rules.

(Q4) *Rule for Unbracketing Lines:* At any stage of the game, the inquirer may unbracket any answer or premise bracketed at the time. Then all the later brackets caused by this line having been bracketed can also be removed.

The reason why we can get along with such simple rules is an important one. One might perhaps want us to formulate rules as to when an inquirer should bracket an answer. Such rules would nevertheless be strategic rules rather than what can be called definitory ones. Such rules do not deal merely with what moves a player may make; they deal with the better and worse ways of applying the definitory rules of a game. It would be a serious mistake to try to incorporate such strategic rules into the definitory rules of one's game.

The detailed study of the strategic rules of bracketing and unbracketing nevertheless belongs in any case to epistemology and to the philosophy of science rather than logic. In practice, such strategic rules of bracketing and otherwise evaluating an oracle's answers operate by means of evaluations of the reliability of the oracle in question. Conversely the main basis of such evaluations of oracles is often their previous track record as answering questions. As an extreme case, an oracle's answer can have two kinds of impact on the reasoner's thinking. It provides prima facie information whose reliability nevertheless has to be tested. But one such possible partial test is the answer itself, in that it can sometimes tell something about the oracle in question. For instance, it can show how well informed the oracle is and how well the new answer coheres with the rest of the inquirers relevant knowledge, which can provide some information about the oracle's reliability. Thus our interrogative logic provides us with an interesting vantage point which is likely to throw some light on the fascinating problem area of cognitive fallacies. (See Piatelli-Palmarini 1994 and the literature referred to there.) We will nevertheless not try to indulge in a full-scale discussion of these matters here. In this article, we will instead study more closely the logics of pure interrogative inquiry where all answers are true and known to be true.

9. INTERROGATIVE REASONING AND IDENTIFIABILITY

One of the many interesting applications of the ideas of interrogative logic concerns the notion of identifiability. (Cf. here Hintikka 1991.) Let T be a theory in a language L, and let P be a one-place predicate symbol occurring in T. We say that P is *explicitly definable* on the basis of T if and only if

$$(9.1) \qquad T \vdash (\forall x)[P(x) \leftrightarrow D[x]]$$

where D(x) is a formula of L such that P does not occur in it and x is the only free variable in it.

It is possible to generalize the notion of definability by using our interrogative logic. We call this generalized notion *identifiability*.

We say that P is *explicitly identifiable* on the basis of T in a model M if and only if

$$M{:}T \vdash (\forall x)[P(x) \leftrightarrow D[x,a_1,\ldots,a_n]]$$

where $D[x,y_1,\ldots,y_n]$ is a formula of L such that P does not occur in it and $x,y_1,\ldots,y_n$ are the only free variables in $D[x,y_1,\ldots,y_n]$. Individuals $a_1,\ldots,a_n$ are from do(M) ($n < \omega$). Auxiliary information may not contain P.

If there are no restrictions on the available answers, then explicit identifiability reduces to definability in the model M with parameters. If n=0, and if the inquirer may not ask any questions, then explicit identifiability reduces to explicit definability. If the inquirer uses questions in the proof of the definition, then we have

$$(9.2) \qquad M{:}T \vdash (\forall x)[P(x) \leftrightarrow D[x]]$$

where D[x] is as in an explicit definition of P. In order to obtain (9.2) we use auxiliary information from the model, so (9.2) is not equivalent with (9.1). Clearly, (9.1) entails (9.2), but not vice versa.

It is possible to define generalizations of other kinds of definitions, for example, we say that P is *piecewise identifiable* on the basis of T in a model M if and only if

$$M:T \vdash \vee_{i=1}^{i=k}(\forall x)[P(x) \leftrightarrow D_i[x,a_1,...,a_n]]$$

where each D_i (i=1,...,k) are as in the case of explicit identifiability.

Let Q be a new one-place predicate symbol. Let T′ be the theory obtained from T by substituting Q for P in T. (The language L(Q) used in T′ is L extended by adding Q to it.) We say that P is *implicitly identifiable* on the basis of T in a model M if and only if

$$M:(T \cup T') \vdash (\forall x)[P(x) \leftrightarrow Q(x)].$$

Auxiliary information may not contain P or Q.

The following theorem is the very basic theorem in the theory of identifiability. (It is closely related to the Covering Law Theorem of sec. 4.) Let C be a sentence of L(Q).

Extended Interpolation Theorem: Assume that T is a consistent theory such that

(i) **A**:T ⊢ C and
(ii) not: T ⊢ C.

Then there is an interpolant $I[a_1,...,a_n]$ such that

(a) each nonlogical constant of $I[a_1,...,a_n]$ occurs in both T and C, except for a finite number of individual constants $a_1,...,a_n$,
(b) **A**:T ⊢ $I[a_1,...,a_n]$,
(c) $I[a_1,...,a_n]$ ⊢ C.

As a consequence of the Extended Interpolation Theorem we obtain the following Extended Beth's Theorem (cf. Beth 1953):

Extended Beth's Theorem: P is explicitly identifiable on the basis of T in M if and only if P is implicitly identifiable on the basis of T in M.

These theorems are proved in Hintikka and Harris (1988). Extended Beth's Theorem helps us to use the techniques of interrogative logic in the theory of identifiability just as Beth's theorem helps us in the theory of definitions.

We obtain a generalized notion of identifiability by stipulating that P is identifiable on the basis of T in every model of T. There are two different interpretations on this generalized notion: the defining formula $D[x,y_1,...,y_n]$ is either the same formula in every model of T or it can be different in different models. By means of this generalized notion of identifiability we can show some interesting interconnections between the notions of definability and identifiability. (See Hintikka 1991.)

The framework we have formulated enables us to make several useful general conceptual distinctions. For instance, we can now clear up some of the confusions that have affected the earlier discussion of the role of theoretical concepts in scientific theories. What needs to be done here is to distinguish the question as to which terms can occur in the answers that the oracle (or the oracles) will provide from the question which concepts occurring in a given theory T are identifiable. These are distinctly different questions. For instance, only the latter, but not the former, depends on the background theory. The former question might be called a question of observability, the latter a question of identifiability.

In practice, questions of observability seldom, if ever, have completely general answers. Even the most typical observable concepts, such as distance, are observable only over a certain range of values. Very large and very small distances have to be determined inferentially, instead of being directly observed. Hence it is questionable whether a sharp concept of observability is very useful in the first place.

In contrast, in questions of identifiability we can look away from the contingent restrictions on nature's answers and concentrate only on the logical structure of the theory T plus the structural restrictions on nature's answers. It is likely to play an important role in the general methodology of science, as it has already played in the methodology of some particular sciences, for instance in econometrics. (Cf. here Hsiao 1983.) For one thing, classical problems of definability in science frequently turn out to pertain to identifiability rather than definability in the strict sense of the word used here.

10. SEMANTICAL GAMES AND INFORMATIONAL INDEPENDENCE

The simple logic of interrogation so far outlined is in many ways instructive as an approximation to the full explicit logic of questioning. Nevertheless it must be refined in several ways. Perhaps the most important weakness it has is that it can only be used to study interrogative arguments with a predetermined ultimate conclusion. This does not deprive it of all interest. It can be used to study the confirmation and explanation of conclusions. It can be used for the purposes of answering propositional questions by means of interrogative inquiry. Suppose you want to answer by means of interrogative inquiry the question whose presupposition is

$$(S_1 \vee S_2 \vee \ldots \vee S_n)$$

starting from the initial premises T. All you need to do for the purpose is to try to construct n parallel *tableaux* which start from the sequents

$T \rightarrow S_i$

(i=1, ..., n). This treatment applies *a fortiori* to yes-no questions, but it cannot be extended to questions where a disjunction is independent of K but dependent on one or more universal quantifiers. Most importantly, the simple form of interrogative inquiry so far studied cannot be used for the purpose of answering wh-questions by means of interrogative reasoning. Yet such processes of answering "big" (principal) questions by putting a number of "small" (operative) questions to one's oracles is the most common and the most important use of interrogative reasoning. This double role of questions is instantiated in Plato's dialogues where Socrates examines an overarching question ("What is knowledge?," "What is piety?" etc.) by putting a number of smaller questions to his interlocutor. Likewise, Aristotle was aware of the distinction between principal and operative questions and warned us against confusing the two.

Furthermore, the simple form of our interrogative logic does not explain the relationship between questions, their presuppositions, and their (conclusive) answers. For not any singular term of the appropriate sort can qualify as an answer in the sense of a conclusive answer, that is to say, can do the job it is semantically speaking calculated to do. For instance, if I ask,

(10.1) Who murdered Roger Ackroyd? or
(10.2) Who is Lord Avon?

you are not providing a conclusive answer if you respond by saying

(10.3) Whoever murdered him or
(10.4) Lord Avon, who else?

Both these difficulties can be overcome in the same way. What we have to do is to construe interrogative inquiry as a search of knowledge and not just as an exercise in argumentative bridge-building. This means enrolling epistemic logic to our service. The initial premises are now expressed by the inquirer by "I know that S_i," where S_i is one of the propositions that previously was called an initial premise. They will be abbreviated by $K_I S_i$. Since only one inquirer is considered in this paper, we will omit in the sequel the subscript to K which indicates the knower. An answer to a question will also have the form KA, as will have the ultimate conclusion.

In order to develop a viable epistemic logic that could help us here, we first have to formulate a semantics for epistemic statements. This is best done (for reasons which soon will become clear) in game-theoretic form. In the simplest case (which is the only one dealt with explicitly here) we are dealing only with the case of a single knower. Hence we can simply use K without a subscript. Then the interpretation of an epistemic sentence is with respect to a model system which is a set Ω of models of the (nonepistemic part of the) underlying language, together with a two-place relation defined on this set. This relation is called the epistemic alternativeness relation. Intuitively, given a model ("possible world") M, the epistemic alternatives to M are all the worlds that are compatible with everything the knower in question knows in M.

Then the rules for semantical games are obvious. Such a game begins in the model in which the given sentence S is evaluated. At each stage, the players are considering a sentence and a model.

(R.&) $G((S_1 \& S_2))$ begins with the falsifier's choice of S_i (i = 1 or 2). The game is then continued as in $G(S_i)$.

(R.∨) $G((S_1 \vee S_2))$ begins with the verifier's choice of S_i (i = 1 or 2). The game is then continued as in $G(S_i)$.

(R.A) $G((\forall x)S[x])$ begins with the choice by the falsifier of a member, say b, of do(M). (If this individual does not have a name in the underlying language L, it is given one which is added to L.) The rest of the game is as in G(S[b]), where S[b] is the sentence obtained from $(\forall x)S[x]$ by omitting the quantifier and by replacing all occurrences of the variable bound to it by "b."

(R.E) Likewise, except that choice is made by the verifier.

(R.~) G(~S) begins with an exchange of roles of the two players, as defined by the game rules (including rules for winning and losing). The rest of the game is otherwise as in G(S).

Semantical games of this sort must be distinguished sharply from the epistemic questioning games. The distinction and its significance is spelled out in Hintikka (1996 (b) chapter 2, and 1995).

11. A NEW EPISTEMIC LOGIC FOR DIFFERENT TYPES OF KNOWLEDGE

The game-theoretical nature of the semantics we have developed enables us to take the crucial step needed to develop a viable logic of knowledge, questions and answers. It is to allow the moves connected with disjunctions and existential quantifiers (when made by the initial verifier, ditto for moves for conjunctions and universal quantifiers, when likewise made by the initial verifier) informationally independent of a sentence-initial K. (Cf. Hintikka 1996 (b), chapters 4 and 11.) Such independence provides the precise logico-semantical realization of the question ingredient present in natural languages. We will indicate such independence by a slash notation, writing e.g. $(\exists x/K)$ instead of $(\exists x)$

and $(\vee/K)$ instead of $\vee$.

By means of this notation we can express the logical forms of all the different types of knowledge.

Here we focus (for reasons of space) on what might be called nonmultiple (single-target) knowledge, excluding such kinds of knowledge as involve the specification of two different items. For instance, sentences like the following will not be directly treated here:

Mary knows to whom John gave what

even though it can be represented (on its different readings) in our treatment as

$$K_{Mary}(\exists x/K_{Mary})(\exists y/K_{Mary})(\text{John gave y to x})$$

or as

$$K_{Mary}(\forall y)((\exists x)(\text{John gave y to x}) \supset (\exists x/K_{Mary})(\text{John gave y to x})).$$

Indeed, our discussion of nonmultiple questions can be without any problems extended to multiple questions. Their theory has been studied (in a framework not involving informational independence) in Hintikka 1976.

The kinds of knowledge considered here are represented by *knowledge statements*. They are of the form KS where S is an ordinary first-order statement (in negation normal form) except that one of the existential quantifiers $(\exists x)$ has been replaced by $(\exists x/K)$ or one of the disjunctions $(S_1 \vee S_2)$ has been replaced by $(S_1 (\vee/K) S_2)$. Also, K_I is shortened to K. Our treatment of this case can be extended to knowledge statements containing several slashes.

The following list of knowledge statements and their formalizations illustrates the power and flexibility of our "independence-friendly" slash notation.

(11.1a) I know that S

(11.1b) KS

(11.2a) I know whether S

(11.2b) $K(S (\vee/K) \sim S)$

(11.3a) I know whether S_1 or S_2

(11.3b) $K(S_1 (\vee/K) S_2)$

(11.4a) I know who (say x) is such that S[x]

(11.4b) $K(\exists x/K)S[x]$

(11.5a) I know to whom each person bears the relation S[x,y]

(11.5b) $K(\forall x)(\exists y/K)S[x,y]$

(11.6a) I know of each person (say x) whether he or she satisfies the condition $S_1[x]$ or $S_2[x]$

(11.6b) $K(\forall x)(S_1[x] (\vee/K) S_2[x])$.

Of these, (11.1)-(11.4) have an equivalent slash-free first-order representation, while (11.5)-(11.6) do not.

These different kinds of constructions illustrate the role of informational independence in knowledge statements. Consider, for instance, a statement of the form

(11.7) $K(\exists x/K)M(x,r)$

as distinguished from

(11.8) $K(\exists x)M(x,r)$

where M(x,r) might for instance be "x murdered Roger Ackroyd." Then (11.7) says merely that I know that someone murdered Roger Ackroyd. In contrast, (11.8) says that in each world compatible with my knowledge there is an individual *chosen independently of the world in question* and hence the same in all of them. In other words, there is an individual such that my knowledge rules out all possibilities that that individual did not murder Roger Ackroyd. And this obviously means that *I know who* murdered Roger Ackroyd.

In the case of (11.7) the same effect could be obtained by operator ordering in that (11.7) is logically equivalent with $(\exists x)KM(x,r)$. But an existential quantifier, even though independent of K, may nevertheless depend on some intermediate universal quantifier, as in

(11.9) $K(\forall x)(\exists y/K)S[x,y]$.

Then the knowledge statement does not have any slash-free equivalent.

12. AN EPISTEMIC LOGIC FOR QUESTIONS AND ANSWERS

These different kinds of knowledge statements (with the exception of simple *knows that* statements) can operate as *desiderata* of so many kinds of questions. The desideratum of a question specifies the epistemic state that a conclusive answer to it is calculated to bring about (on the assumption that the presupposition of the question is true) in a typical information-acquiring use of questions. The direct questions whose desiderata (11.2)-(11.5) are can accordingly be expressed as follows:

(12.1c) Is it the case that S?

(12.2c) Is it the case that S_1 or S_2?

(12.3c) Who (say x) is such that S[x]?

(12.4c) To whom does each person bear the relation S[x,y]?

(12.5c) Does each person (say x) satisfy the condition $S_1[x]$ or $S_2[x]$?

Like (11.1a)-(11.6a), (12.1c)-(12.5c) are not idiomatic English sentences. However, it is easy to construct such sentences with the same logical form as (11.1a)-(11.6a) and (12.1c)-(12.5c).

Our slash notation enables us to formulate all the basic concepts concerning questions and answers. For instance, the *presupposition* of a question can be obtained from its desideratum simply by omitting all the slashes, for instance replacing $(\exists x/K)$ by $(\exists x)$. Thus the presuppositions of (12.1c)-(12.5c) are:

(12.1d) $K(S \vee \sim S)$

(12.2d) $K(S_1 \vee S_2)$

(12.3d) $K(\exists x)S[x]$

(12.4d) $K(\forall x)(\exists y)S[x,y]$

(12.5d) $K(\forall x)(S_1[x] \vee S_2[x])$.

A *reply* results from the desideratum by omitting the slashed quantifiers and replacing the variables bound to one of them by an individual or function constant of the appropriate kind. For instance, the following are the logical forms of replies to (12.1c)-(12.5c):

(12.1e) $KS, K{\sim}S$

(12.2e) KS_1, KS_2

(12.3e) $KS[b]$

(12.4e) $K(\forall x)S[x,f(x)]$

(12.5e) $K(\forall x)((S_1[x] \ \& \ f(x)=0) \vee (S_2[x] \ \& \ f(x)\neq 0))$.

A reply is a *conclusive answer* if and only if the identity of the referents of all the substituting individual and function constants is known. For instance, the conclusiveness conditions for (12.3e)-(12.5e) can be expressed as follows:

(12.3f) $K(\exists x/K)(b=x)$

(12.4f) = (12.5f) $K(\forall x)(\exists y/K)(f(x)=y)$.

This is logically equivalent with either of the following

$$K(\exists g/K)(\forall x)(f(x)=g(x))$$
$$(\exists g)K(\forall x)(f(x)=g(x))$$

which show the parallelism between (12.3f) and (12.4f).

It can be shown that a reply to a question logically implies its desideratum if the corresponding conclusiveness condition is satisfied. This shows that the conclusiveness conditions really do the job they were calculated to do, which is to characterize the relationship of a question to its (conclusive) answers. The weak point in all the earlier approaches to the logic and semantics of questions and answers (cf., e.g. Collingwood 1933, Belnap and Steel 1976, Harrah 1984, Bromberger 1992, Wisniewski 1995) is that this fundamental relationship between questions and their (conclusive) answers is not done justice to in them.

Replies to a given question which do not satisfy the conclusiveness condition can convey to the questioner some useful information, albeit not all the information needed to satisfy the questioner. Such replies can be called *partial answers*. They deserve (and need) a separate study.

13. CONCLUSIVENESS CONDITIONS AS CODIFYING CONCEPTUAL KNOWLEDGE

Thus we have reached a deeper understanding of the nature of answers to questions in the interrogative inquiry — and of the nature of questions and answers in general. One striking fact here is that two different things are required for a conclusive answer to a question, on the one hand a reply to it and on the other hand the conclusiveness condition for that reply. In general, these two do not always come from same source or represent the same kind of information. For instance, consider an experimental question whose desideratum is of the form

(13.1) $K(\forall x)(\exists y/K)S[x,y]$.

It is a question concerning how the variable y depends on the variable x in S[x,y]. A reply to such a question is of the form

(13.2) $K(\forall x)S[x,g(x)]$.

Here we can think of the function g(x) as being given extensionally as a class of ordered pairs of values of the controlled variable x and corresponding measured values of the observed variable y=g(x). Such a way of being given does not yet tell the inquirer which mathematical function g(x) is. This additional information is what is expressed by the conclusiveness condition

(13.3) $K(\forall x)(\exists y/K)(g(x)=y)$

which may also be expressed as

(13.4) $K(\exists f)(\forall x)(g(x)=f(x))$

or even

(13.5) $K(\exists f)(g=f)$.

This shows that what the conclusiveness condition (13.3) says is that the inquirer knows which function the function g(x) is that figures in the given reply, i.e. knows the (mathematical) identity of this function.

But if so, it is clear that the information which (13.3) yields to the inquirer is not given to him or her through the experiment itself. It constitutes *a priori* mathematical knowledge which comes to him from an entirely different source. As we might illustrate the difference, in order to obtain the reply (13.2) a theoretical physicist has to appeal to his experimentalist colleagues, but in order to reach the conclusiveness condition (13.3) he or she has to sit with mathematicians and commit *a priori* reasoning.

Hence it is reasonable to require of an interrogative step only that a reply to the question in question is given by the oracle. What is needed then is another oracle that answers questions of identification, i.e. questions concerning which objects of different logical types can be considered as being known by the inquirer. We might call this the identification oracle. The need to postulate such an oracle as a part of the interrogative model shows that *a priori* truth does not enter empirical inquiry only through the logical inference steps of inquiry. An independent source of such truths is in principle intrinsic to all reasoning and inquiry. We will call such a source of answers an *a priori* oracle in contradistinction to ordinary *a posteriori* ones, and we will apply the same terms to the steps of consulting the one or the other kind of oracle.

However, a further explanation is in order here. In one respect, the term *a priori* knowledge is misleading. If we look at the role of conclusiveness conditions in interrogative inquiry, we can see that they have to be reached in principle in the same way as all the other conclusions which serve as stepping-stones to further ones, viz. by means of the same interrogative logic. They need not be given, as initial premises or in any other way. In this respect, it may very well be misleading to call them *a priori*. In general, there is no reason not to think of mathematical and other conceptual knowledge as also being reached by means of interrogative reasoning, not unlike ordinary empirical knowledge. However, this does not eliminate a distinction between the two in that they rely on the testimony of different kinds of oracles.

One further comment is needed here. Usually it does not make sense to say that the choices of the values of two adjacent universal quantifiers are dependent or independent of each other. Hence there seems to be no difference between $K(\forall x)$ and $(\forall x)K$, for K is in effect a universal quantifier ranging over epistemic alternatives. By the same token, the distinction between $K(\forall x)$ and $K(\forall x/K)$ might seem to be without difference. But a second thought shows that the ranges of universal quantifiers inside and outside the scope of K are different. In $K(\forall x)$ the values are the individuals existing in the world whose choice is governed by K, whereas in $(\forall x)K$ and $K(\forall x/K)$ the choice is from the individuals that are well defined for all the epistemic alternatives. Hence we need also slashed quantifiers of the form $(\forall x/K)$.

14. THE TWO ROLES OF QUESTIONS IN INTERROGATIVE LOGIC

By means of the epistemic logic and the theory of questions and answers so developed, we can study the logic of interrogative inquiry in greater detail than before. We will call the result epistemic interrogative logic. The first main change is that the initial premises and the ultimate conclusion (the statement to be established) are knowledge statements. Our task is then to reformulate the rules of interrogative inquiry which now involve an epistemic element and then study the resulting epistemic interrogative logic.

The principal novelty here is the possibility of using the desideratum of a question (i.e. of the principal question of an inquiry) as the ultimate conclusion. Then the interrogative argument leading to that conclusion (assuming that the game *tableau* is closed) is tantamount to answering the principal question by means of questioning.

In other words in such an interrogative inquiry the principal question is answered with the help of "small" or operational questions put to the oracle. This is eminently in keeping with the historical precedents of our interrogative logic. The main difference is that now we have as our disposal an explicit theory of questions, answers and the question-answer relationship.

In this way we can bring our interrogative logic to bear on problem-solving. In some ways, the extension (epistemic version) of our interrogative logic can be thought of as a general logic of problem-solving. This approach is especially natural in that our treatment will automatically involve a treatment of the different strategies of problem-solving. In such a treatment, the Strategy Theorem of section 4 above (together with its extensions) is obviously going to play a crucial role.

The other main advantage offered by an epistemic approach to interrogative inquiry is a closer analysis of the nature of the answers that are used in interrogative inquiry as new premises. To take an example, in the first-order approximation to a logic of interrogative inquiry, an answer to a question whose presupposition is

(14.1) $(\forall x)(\exists y)S[x,y]$

was of the form

(14.2) $(\forall x)S[x,g(x)]$.

It does not seem to matter very much to the logical relations of these two statements whether we prefix both of them by K or not. But in a deeper perspective we cannot restrict our attention any longer to (14.1)-(14.2) or to their epistemic versions

(14.3) $K(\forall x)(\exists y)S[x,y]$

(14.4) $K(\forall x)S[x,g(x)]$.

For we know now, on the basis of the analysis of questions and answers offered in section 12 above, that the function g(x) does not automatically yield a conclusive answer, i.e., that (14.4) is not always a conclusive answer to the question whose presupposition is (14.3). What is also needed is the conclusiveness condition

(14.5) $K(\forall x)(\exists y/K)(g(x) = y)$.

This conclusiveness condition could also be written in the following forms:

(14.6) $K(\exists f/K)(\forall x)(g(x) = f(x))$

(14.7) $(\exists f)K(\forall x)(g(x) = f(x))$

(14.8) $(\exists f)K(g = f)$.

This shows the intuitive import of the conclusiveness condition in this case. What (14.4) says is that the function g(x) is of the kind desired as an answer. What (14.8) says, i.e., what the conclusiveness condition says is that it is known which function g(x) is.

Hence in the nonepistemic (first-order) formulation of the logic of interrogative inquiry we had to assume of each available answer that the relevant conclusiveness condition is satisfied. This impoverishes one's interrogative logic greatly. For one thing, the reply (4) and the conclusiveness condition (14.5) might come from different sources. For instance, the function g(x) might be given (in extension, so to speak) by a scientific experiment. Such a function would not yield a satisfactory answer unless or until it is known which mathematical function we are dealing with. And the information that makes that condition true is likelier to come from mathematics than from a physical experiment.

Thus the epistemic approach opens up the possibility of a sharper analysis of the role of answers and of their conclusiveness conditions in interrogative inquiry. Indeed, our comments on (14.1)-(14.8) concern essentially what was called above in section 5 the AE-case, which includes the logic of experimental inquiry. A moment's thought shows that without taking the epistemic element into account we cannot even do justice to the presuppositions of such questions, let alone to the conclusiveness conditions of their answers.

15. EPISTEMIC RULES FOR INTERROGATIVE LOGIC

But what do the new, improved rules for interrogative inquiry look like? At first sight it might seem impossible to formulate a complete set of rules for the purpose. This impossibility is suggested by independence-friendly first-order logic, for which a complete set of rules of proof cannot be formulated, even though there is complete set of rules of disproof, that is, for proving logical inconsistency. This is connected with the fact that IF first-order languages are not closed with respect to contradictory negation. (See here Hintikka 1996 (b), chapters 3-4, 6.) Hence we cannot show that $(F \supset G)$ is valid by showing that there is a model in which F is true but G not true, for G's not being true cannot be expressed by any IF first-order formula. The simplest sentences whose contradictory negations fail to be expressible in an IF first-order language are the Henkin quantifier formulas, i.e., formulas of the form

(15.1) $(\forall t)(\exists x)(\forall y)(\exists z/\forall t)S[t,x,y,z]$.

Now knowledge statements may seem to exhibit this same quantifier structure when we realize that the knowledge operator K is a disguised universal quantifier. For instance, knowledge statements of the following form may seem to have the same difficulties associated with them as Henkin quantifier sentences the IF first-order logic:

(15.2) $K(\exists x)(\forall y)(\exists z/K)S[x,y,z]$.

Luckily this first impression turns out to be incorrect. The reason is that the "quantifier" K is in a certain sense nonstandard. In order to see this subtle point consider a knowledge statement of the form

(15.3) $K(\forall x)(\exists y/K)S[x,y]$.

This statement is true if and only if

(15.4) $(\exists f)K(\forall x)S[x,f(x)]$.

But if f is considered as a function of two variables, x and k, the relevant epistemic alternative f(x,k) cannot be chosen arbitrarily. It has to be a function of the *individual* x which is well defined for all the epistemic alternatives. It follows that whenever the value of x in two alternatives is the same, the value of f(x) in these two worlds, must also coincide. This restricts the set of the functions f over which the function quantifier $(\exists f)$ ranges and turns this quantifier in effect into a first-order quantifier ranging over the world lines of individuals well-defined in all alternatives. For this reason we may hope to be able to formulate a complete set of rules for interrogative inquiry, in spite of the presence of the phenomenon of informational independence.

In formulating the new rules for interrogative logic we will assume throughout that the formulas we are dealing with are in the negation normal form.

The rules for interrogative logic involving epistemic notions can be formulated in terms of what we will call k-sequents. We can think of them as ordinary sequents in which each formula is thought of as containing a suppressed initial K and in which there is initially only one formula on the right side. In terms of such k-sequents, we can formulate the following rules:

(L.&), (L.∨f), (R.&f), (R.∨), (L.E), (R.A), (L.A), (R.E) as before.

(L.E/K) $$\frac{\Gamma, S_0[(\exists x/K)S_1[x]] \rightarrow \Delta}{\Gamma, S_0[S_1[f(y_1,\ldots,y_i)]],\ (\forall y_1)\ldots(\forall y_i)(\exists x/K)(f(y_1,\ldots,y_i) = x) \rightarrow \Delta.}$$

Here f is a new function symbol and $(\forall y_1),\ldots,(\forall y_i)$ are all the universal quantifiers within the scope of which $(\exists x/K)$ occurs in $S_0[(\exists x/K)S_1[x]]$.

(R.E/K) $$\frac{\Gamma,(\exists x/K)(t = x) \rightarrow \Delta, S_0[(\exists x/K)S_1[x]]}{\Gamma,(\exists x/K)(t = x) \rightarrow \Delta, S_0[(\exists x/K)S_1[x]], S_0[S_1[t]]}$$

where t is any constant term.

In our epistemic version of interrogative logic, the counterpart to the tautology introduction rule will be the following

(L.taut) $$\frac{\Gamma \rightarrow \Delta}{\Gamma, K(S \vee \sim S) \rightarrow \Delta}$$

where S is any formula in the underlying language.

The rule (R.cont) cannot be formulated quite analogously, because there must not be more than one formula on the right-hand side of the sequent:

$$\frac{\Delta \rightarrow KS_1}{\Gamma \rightarrow K(S_1 \vee (S_2 \,\&\, \sim S_2))}$$

where S_2 is any formula in the appropriate language.

We need also rules for independent disjunctions.

(L.∨/K) $$\frac{\Gamma, S_0[S_1 \,(\vee/K)\, S_2] \rightarrow \Delta}{\Gamma, S_0[(S_1 \,\&\, f(y_1,\ldots,y_i) = 0) \vee (S_2 \,\&\, f(y_1,\ldots,y_i) \neq 0)] \,\&\, (\forall y_1)\ldots(\forall y_i)(\exists x/K)(f(y_1,\ldots,y_i) = x) \rightarrow \Delta}$$

where f is a new function symbol and $(\forall y_1),\ldots,(\forall y_i)$ are all the universal quantifiers within the scope of which $(S_1 \,(\vee/K)\, S_2)$ occurs in S_0.

(R.∨/K) $$\frac{\Gamma \rightarrow \Delta, S_0[S_1 \,(\vee/K)\, S_2]}{\Gamma \rightarrow \Delta, S_0[(\exists x/K)((S_1 \,\&\, x = 0) \vee (S_2 \,\&\, x \neq 0))].}$$

16. ON THE ROLE OF CONCLUSIVENESS CONDITIONS

The way these rules operate can be illustrated by showing how a reply and the conclusiveness condition in some selected cases imply the desideratum of a question. In the following proofs, we are leaving all appeals to structural rules (i.e. rules for permuting members of sequents) tacit. The sequents involved are of course k-sequents.

(i) $(\forall x)S[x,g(x)], (\forall x)(\exists y/K)(g(x) = y) \rightarrow (\forall x)(\exists y/K)S[x,y]$

(Cf. here (14.3)-(14.5).)

(ii) $(\forall x)S[x,g(x)], (\forall x)(\exists y/K)(g(x) = y) \rightarrow (\exists y/K)S[b,y]$

from (i) by (R.A).

(iii) $(\forall x)S[x,g(x)], (\forall x)(\exists y/K)(g(x) = y), S[b,g(b)], (\exists y/K)(g(b) = y) \rightarrow (\exists y/K)S[b,y]$

from (ii) by two application of (L.A) (and of structural rules).

(iv) $(\forall x)S[x,g(x)], (\forall x)(\exists y/K)(g(x)=y), S[b,g(b)], (g(b)=d), (\exists y/K)(d=y) \rightarrow \exists y/K)S[b,y]$

from (iii) by (L.E/K).

(v) $(\forall x)S[x,g(x)],(\forall x)(\exists y/K)(g(x)=y), S[b,g(b)], g(b)=d, (\exists y/K)(d = y) \rightarrow (\exists/K)S[b,y], S[b,d]$

from (iv) by (R.E/K).

(vi) $(\forall x)S[x,g(x)], (\forall y)(\exists y/K)(g(x) = y), S[b,g(b)], g(b) = d, (\exists y/K)(d = y), S[b,d] \rightarrow (\exists y/K)S[b,y], S[b,d]$

from (v) by (L.=). Here (vi) is a closure sequent.

Likewise the following k-sequent can be proved.

(i) $(\exists x)(\forall y)S[x,y,g(y)], (\forall x)(\exists y/K)(g(x) = y) \rightarrow (\exists x)(\forall y)(\exists z/K)S[x,y,z]$.

The proof might run as follows:

(ii) $(\forall y)S[a,y,g(y)], (\forall x)(\exists y/K)(g(x) = y) \rightarrow (\exists x)(\forall y)(\exists z/K)S[x,y,z]$

from (i) by (L.E) (and structural rules).

(iii) $(\forall y)S[a,y,g(y)],(\forall x)(\exists y/K)(g(x)=y) \rightarrow (\exists x)(\forall y)(\exists z/K)S[x,y,z], (\forall y)(\exists z/K)S[a,y,z]$

from (ii) by (R.E).

(iv) $(\forall y)S[a,y,g(y)], (\forall x)(\exists y/K)(g(x) = y) \rightarrow (\exists x)(\forall y)(\exists z/K)S[x,y,z], (\exists z/K)S[a,b,z]$

from (iii) by (R.A).

(v) $(\forall y)S[a,y,g(y)], (\forall x)(\exists y/K)(g(x)=y), S[a,b,g(b)], (\exists y/K)(g(b)=y) \rightarrow$
$(\exists x)(\forall y)(\exists z/K), S[x,y,z], (\exists z/K)S[a,b,z]$

from (iv) by (L.A) (and of structural rules).

(vi) $(\forall y)S[a,y,g(y)], (\forall x)(\exists y/K)(g(x)=y), S[a,b,g(b)], (\exists y/K)(g(b)=y) \rightarrow$
$(\exists x)(\forall y)(\exists z/K)\, S[x,y,z], (\exists z/K)S[a,b,z], S[a,b,g(b)]$

from (v) by (R.E/K). Here (vi) closes the tableau construction and thereby shows that (i) is valid.

Thus an implication from a reply and its conclusiveness condition to the desideratum of a question holds also when the conclusion prima facie has a Henkin quantifier form. This is an indication of the fact mentioned earlier, viz. that the presence of Henkin-type prefixes (as in (15.2) above) in epistemic logic does not prevent the formulation of a complete sets of rules for interrogative logic.

17. REPLIES AND ANSWERS IN INTERROGATIVE REASONING

The question-answer relationship codified by the conclusiveness condition is the crucial novelty of the approach to the logic of questions and answers ("erotetic logic") presented here. Most of the other approaches are reduced to labeling every response to a question "answer" which provides the questioner with some relevant information. It is easy to find a wealth of examples of such responses which nevertheless do not satisfy the questioner but instead prompt further questions.

It is nevertheless important to be clear precisely what role the requirement of conclusiveness plays in the questions and answers that figure in interrogative reasoning. In sec. 14 it was pointed out as a weakness of the simplified nonepistemic version of interrogative logic (secs. 1-9 above) that oracles' responses must all be assumed to be conclusive answers. A stipulation to this effect was implicitly made in the original question rule. This would nevertheless be unrealistic in many applications of our interrogative logic. Now we are in a position to see how the question rule can be made more realistic. What often, perhaps typically, happens in ordinary discourse is that a reply and the corresponding conclusiveness condition are established independently of each other. Only when they are put together do they imply a conclusive answer to a question. We must clearly allow for such a procedure also in our interrogative logic. That means that in the question rule an oracle's response can be a reply and not always a conclusive answer. We can — and perhaps should — require that if an oracle can give a conclusive answer, it will do so. But when there is no conclusive answer to a given question among the ones an oracle can yield, we can allow the oracle to respond by giving a reply to it in our technical sense. The crucially important role of conclusiveness conditions in a systematic theory of questions, answers, and interrogative logic does not mean that an oracle's responses always have to satisfy them. Some requirements of informativeness must nevertheless be imposed on an oracle's replies, in order to avoid the possibility that an obstreperous oracle can frustrate an interrogative argument altogether. We might even try to impose a requirement of maximal informativeness on oracles' replies. In particular, it is in keeping with the spirit of most applications to require an oracle who is answering basic questions of identification to give maximally informative replies. These are questions whose desideratum is of the form

$$K(\forall y_1)...(\forall y_i)(\exists x/K)(f(y_1,...,y_i) = x)$$

if the oracle can give a conclusive answer, it will give one as a reply.

With these modifications, the question rule allows more realistic applications of our interrogative logic than before. Among other things, as was noted, our new rule makes it possible that the inquirer should receive a reply to a question and the corresponding conclusiveness condition separately, perhaps as replies to different questions. This is realistic. In many situations, establishing the conclusiveness condition of a reply is the responsibility of the inquirer, not of the answerer. For instance, it is unrealistic to expect nature to yield more than replies to questions. The inquirer has to gather the corresponding conclusiveness condition from other sources.

18. SOME MAIN RESULTS OF EPISTEMIC INTERROGATIVE LOGIC

For epistemic interrogative logic, we can establish counterparts to the four basic metatheorems formulated in section 4 above.

Completeness Theorem: If a knowledge statement KS is logically implied by Γ plus the set of all answers available, then

$$\Gamma \rightarrow KS$$

can be proved in the extended epistemic interrogative logic.

Yes-No Theorem: In extended epistemic interrogative logic, if a sequent

$$\Gamma \rightarrow KS$$

can be established interrogatively, then it can also be established by using only yes-no questions.

These two metatheoretical results are fairly straightforward extensions of the earlier ones. Their proofs do not prompt any surprises.

The Covering Law Theorem for the nonmodal interrogative logic relied on Craig's interpolation theorem for first-order logic. In order to establish a Covering Law Theorem for interrogative logic, one first has to establish an interpolation theorem for it. Here we only register the fact that such a theorem can be proved, but will not try to do so.

By means of a suitable interpolation theorem we can prove a Covering Law Theorem.

Covering Law Theorem (special case): Assume that the sequent

$$\Gamma \rightarrow KP(b)$$

can be established by means of answers that do not contain P while b does not occur in Γ. Assume also that Γ does not entail KP(b) and that the set of available answers is consistent. Then there is a formula H[x] satisfying the following conditions:

(i) $\Gamma \rightarrow K(\forall x)(H[x] \rightarrow P(x))$

is logically valid;

(ii) $\rightarrow KH[b]$

can be established interrogatively;

(iii) Neither P not b occurs in H[x];

(iv) All nonlogical constants of H[x] occur both in Γ and in the available answers, except for a finite number of individual and/or function constants introduced in the course of the interrogative argument by answers to the inquirer's question.

We will not present an explicit proof of this theorem here or a proof of any of its generalizations. Such a proof would require a fuller discussion of the model theory of epistemic logic than we can undertake here. It is nevertheless in order to indicate the underlying reason why the epistemic version of the Covering Law Theorem holds. It was seen in section 4 above that the Covering Law Theorem holds when we can disregard the epistemic element. Now the *knows that* operator K is in effect merely a universal quantifier ranging over alternative models. Accordingly, it does not introduce anything that could violate the Covering Law Theorem. The operative new elements here are therefore the slashed expressions ($\exists$x/K) and ($\vee$/K). And they can cause havoc in the interpolation theorem (which underlies the Covering Law Theorem) in epistemic interrogative logic only when they occur in the consequent S_2 of the conditional $(S_1 \supset S_2)$ to be interpolated. The reason is that the task of interpolation means finding a separation formula I such that $S_1 \vdash I$, $\neg S_2 \vdash \neg I$ where $\neg$ is contradictory negation. Since I is a formula of ordinary epistemic logic, it does not cause any problems. But it is not clear how the contradictory negation of an epistemic statement can be expressed in general. However, these doubts do not affect the Covering Law Theorem, for there the consequent which is at issue there is KP(b). It does not contain any slashed expressions, and hence its contradictory negation is the same as its normal (strong) negation ~KP(b), which does not take us outside the epistemic logic. Hence there is no obstacle to the use of an interpolation theorem for the purpose of proving the epistemic version of the Covering Law Theorem along roughly the same lines as in the nonepistemic case.

The epistemic Covering Law Theorem can be generalized in the same way as the nonepistemic Covering Law Theorem (special case). (See section 4 above.)

Strategy Theorem: The optimal choice of the question to ask is the same as the optimal choice of the rule (L.E) of (L.∨f) to apply, assuming that the question is answerable.

Of these metatheorems, the first and the third rely on model-theoretical notions like validity. They could be taken for granted in our interrogative logic only as long as the epistemic element was not brought to bear. In the presence of the epistemic operator K, the model theory of knowledge statements need to be spelled out more explicitly.

19. ON THE LOGIC OF EXPLANATORY REASONING

According to what was indicated, in a typical interrogative reasoning the ultimate conclusion, that is, the initial sentence on the right-hand side of the initial sequent is the desideratum of the principal question which the reasoner is trying to answer. What this question is, in other words, what the reasoner is trying to find out, is indicated by the slashed ingredients (∃x/K) and (∨/K) of the desideratum.

But what happens when there are no such ingredients in the ultimate conclusion? In such a case, the reasoner is not trying to discover a new truth, an answer to a question, but to see if (and how) a pre-established conclusion follows interrogatively from given initial premises. What purpose can such an exercise have? What kind of knowledge can one hope to reach by constructing a bridge from the initial premises to the ultimate (but already fixed) conclusion?

On the interpretational level, an answer to the first of these two questions is reasonably clear. Even when we know that a statement is true, we may still want, for a good reason, to know *why* it is true, in other words, we may want to explain its being true Hence an interrogative proof of S from the initial premises T may in many cases be thought of as providing an explanation of why S is true. Why-questions and the kind reasoning needed to answer them thus occupy a natural niche on the map of different types of interrogative reasoning. (Cf. here Hintikka and Halonen 1995.)

The Covering Law Theorem applies to the very case in which the ultimate conclusion does not contain any slashed expressions $(\exists x/K)$ or $(\vee/K)$. The conclusion of the Covering Law Theorem bears in fact some resemblance to what is known as the covering law model of explanation. (See Hempel and Oppenheim 1948, Salmon 1990, Pitt 1988.) There are some marked differences, however. The suggestion we just made was to think of the interrogative proof of the explanandum from the initial premises as providing an explanation. In the strictly construed Hempelian model, transposed to our framework, an explanation is provided by a logical (not interrogative) derivation of the explanandum from the "initial conditions" by the sole means of the covering law. (In our treatment, those initial conditions correspond to nature's answers to the inquirer's questions.) In our treatment of explanation, the covering law does not play any such direct role in explaining. Rather, it looks like a mere by-product of the explanatory argument. In contrast, Hempel assigns in this approach to the covering law the same role as is played in our treatment by the initial premise T (which in scientific context means the background theory). In our view, this is a vicious confusion between the respective roles of the background theory (initial premises) T and of the covering law in explanation. It is not surprising to find in many discussions of the covering law model an unmistakable assimilation of the covering law to the background theory.

Moreover, the covering law itself is available only on certain further conditions, as the Covering Law Theorem shows.

Most importantly, Hempel's covering law model does not provide any explanation of what kind of information it is that the covering law is supposed to yield. This question is obviously a variant of the more general question as to what information explanations of known facts can yield.

An answer can be obtained by having a closer look at the Covering Law Theorem. It suffices to consider the proof of this theorem in the original nonepistemic interrogative logic (see above). The antecedent H[x] of the "covering law"

(19.1) $T \vdash (\forall x)(H[x] \supset P(x))$

is there obtained as the Craigean interpolation formula H between A and $(\wedge T^* \supset P(b))$. It has been shown by Hintikka and Halonen (forthcoming) that (and how) H can be considered as a summary of the argument which leads from A to $(\wedge T^* \supset P(b))$. With certain unimportant exceptions, there obtains a one-to-one correlation between the logical constants in H and the applications of the different rules of interrogative reasoning in the derivation of $(\wedge T^* \supset P(b))$ from A. Hence the interpolation formula H codifies certain information that can be highly nontrivial. This information can be thought of as a blow-by-blow account of how the fact that the "initial conditions" expressed by A hold leads, assuming that the background theory T holds, to the result that P(b). It is the character of H as a summary of the actual connection between A and the explanandum that makes H an integral part of an explanation of P(b). As we may perhaps put it, the argument from A to $(\wedge T^* \supset P(b))$ explains why P(b) holds by showing how the general laws codified in T lead step by step from the initial conditions A to the explanandum P(b), and H provides us with a summary of all these steps in a single sentence.

When the premises of the Covering Law Theorem do not hold, the derivation of $(\wedge T^* \supset P(b))$ from A can still be considered as an explanation of sorts. In this case, it is nevertheless more natural to speak of an explanation *how* rather than of an explanation *why*. Such an account *how* cannot be summarized in any one-line explanation why.

20. EXAMPLES AND COMMENTS

In order to illustrate these remarks, we may consider a sample interrogative argument. In this example,

$$T = \text{initial premise} = (\forall x)(\forall y)(\forall z)((Q(x) \;\&\; R(x,y) \;\&\; R(y,z)) \supset P(z)).$$

Its negation $\sim T$ is $(\exists x)(\exists y)(\exists z)(Q(x) \;\&\; R(x,y) \;\&\; R(y,z) \;\&\; \sim P(z))$.

A = the set of requisite answers to questions = $\{Q(a), R(a,b), R(b,c)\}$.
U = ultimate conclusion = $P(c)$.

Here the antecedent H[x] of the covering law

$$(\forall x)(H[x] \supset P(x))$$

is obtained (as indicated in the proof of the covering law theorem above) as the interpolation formula from the proof of the sequent

$$A \rightarrow (T \supset U)$$

i.e.

$$Q(a), R(a,b), R(b,c) \rightarrow (\forall x)(\forall y)(\forall z)((Q(x) \;\&\; R(x,y) \;\&\; R(y,z)) \supset P(z)) \supset P(c)). \quad (20.1)$$

A (cut-free) proof might run as follows:

$$A \rightarrow \sim T, P(c) \quad (20.2)$$

from (20.1) by (R.∨).

(20.3) $A \rightarrow \sim T, P(c), (\exists y)(\exists z)(Q(a) \ \&\ R(a,y) \ \&\ R(y,z) \ \&\ \sim P(z))$

from (20.2) by (R.E).

(20.4) $A \rightarrow \sim T, P(c), (\exists y)(\exists z)(Q(a) \ \&\ R(a,y) \ \&\ R(yz) \ \&\ \sim P(z)), (\exists z)(Q(a) \ \&\ R(a,b) \ \&\ R(b,z) \ \&\ \sim P(z))$

from (20.3) by (R.E).

(20.5) $A \rightarrow \sim T, P(c), (\exists y)(\exists z)(Q(a) \ \&\ R(a,y) \ \&\ R(y,z) \ \&\ \sim P(z)), (\exists z)(Q(a) \ \&\ R(a,b) \ \&\ R(b,z) \ \&\ \sim P(z)), Q(a) \ \&\ R(a,b) \ \&\ R(b,c) \ \&\ \sim P(c)$

from (20.4) by (R.E).

At this point the argument splits into four branches (*subtableaux*). In the first three (sequents (6)-(8)) the new formula introduced to the right side is Q(a), R(a,b) and R(b,c), respectively. These branches close because the same formula occurs on the left side.

In the fourth case (sequent (9)), the new formula is ~P(c). The branch closes because the formula P(c) also occurs on the right side.

The interpolation formula $I_1[c]$ for (20.1) is obtained as indicated step by step from the argument (1)-(9) as follows:

$I_1[c] = I_2[c]$
$I_2[c] = (\exists x)I_3[x,c]$
$I_3[a,c] = (\exists y)I_4[a,y,c]$
$I_4[a,b,c] = I_5[a,b,c]$
$I_5[a,b,c] = I_6[a] \vee I_7[a,b] \vee I_8[b,c]$
$I_6[a] = Q(a)$
$I_7[a,b] = R(a,b)$
$I_8[b,c] = R(b,c)$
$I_9 = \text{empty}$

Here the existential quantifiers (∃x) and (∃y) making their appearance in I_2 and I_3, respectively, are occasioned by the fact that the substituting constants a and b are imported from the left side to the right one by an application of (R.E).

Putting all the equations I_1 - I_9 together we obtain as $I_1[c] = H[c] =$

$(\exists x)(\exists y)(Q(x) \ \&\ R(x,y) \ \&\ R(y,z)).$

Hence the covering law C is

$$(\forall x)\{(\exists x)(\exists y)(Q(x) \,\&\, R(x,y) \,\&\, R(y,z)) \supset P(z)\}$$

and the "initial condition" H[c] is

$$(Q(a) \,\&\, R(a,b) \,\&\, R(b,c)).$$

Here clearly the two characteristic sequents are provable, viz.

$T \to C$ and
$A \to H[c]$.

It can be seen where the logical constants of H[c] come from. The two existential quantifiers come from the two applications of (R.E) which import the constants a and b, respectively, for the first time from the left side to the right side in steps (3) and (4). The conjunction signs come from the application of (R.$\vee$) in steps (6)-(8). The other steps in (1)-(9) do not introduce any constants into the interpolation formula.

In general, it is seen from the way in which the sequent

$$A \to (T \supset P(c))$$

is proved (so as to give rise to the interpolation formula H[c]) that the interpolation formula is to all practical purposes almost a substitution instance of the initial premise T unless it contains complex quantifier formulas (obtainable as answers to the reasoner's question) which can introduce new constants into the argument by means of (L.E). Unless this is the case, in other words, unless quantificationally complex propositions are available as answers to questions the covering law is little more than an instance of the initial premise. This observation throws some interesting light on the logic of explanatory reasoning. In the theory of explanation, a confusion has been rampant between the roles of the initial premise T (which in scientific explanation typically means the background theory) and the covering law $(\forall x)(H[x] \supset P(x))$ in explanation. What has been seen means that this confusion, too, can be blamed on the Atomistic Postulate. Thus the interrogative logic developed here throws some sharp light on the nature of explanatory reasoning.

21. FALLACIES

Among many other topics in the theory of applied logic, our interrogative approach throws light on the theory of mistakes in reasoning, traditionally known as fallacies. (Cf.

here Hintikka 1987.) Indeed, our theory offers an overview on different kinds of such mistakes. Corresponding to the distinction between logical inference steps and interrogative (questioning) steps, there is a distinction between such fallacies as affect logical inference steps in interrogative inquiry and such as affect questioning steps. Mistakes due to ambiguity fall into the former category, as do such outright misinferences as the mistake known as the fallacy of affirming the consequent. Mistakes in questioning include the so-called fallacy of many questions. In our framework, this mistake amounts to asking a question whose presupposition has not been established. The traditional name of this fallacy is itself fallacious, however. It reflects the idea that the presupposition of the illicit question ought to have been established by a previous question, so that the mistake can be thought of as a misguided attempt to combine two or more questions ("many questions") into one. However, from our rules of interrogative logic (see sections 2 and 15 above) it is seen that the presupposition can equally well be reached by means of a logical inference or even by a series of steps involving both inferences and questions. Hence the intended mistake cannot be thought of simply as an attempt to compress several questions into one.

An especially instructive "fallacy" is the so-called fallacy of begging the question or *petitio principii*. It is in our days usually taken to amount to assuming the conclusion of a logical inference in its premises. As has been pointed out repeatedly, such "assumption" is merely a mark of a valid deductive inference. So where is the fallacy? Here history offers an interesting sidelight on our theory. As Richard Robinson (1971) has shown, originally the intended mistake was to ask ("petition") in an interrogative game the overall principal question instead of trying to answer it by means of a series of "small" or operative questions. The naturalness of this sense of "begging the (wrong) question" is brought out by interrogative logic. A mistake which has usually been taken to be a fallacy in inference turns out to have originally been a fallacy in questioning.

Another distinction between that is made obvious by our framework separates violations of the definitory rules of an interrogative game from violations of its strategic principles. A former kind of violation means that you are not engaged in a honest interrogative enterprise, as little as pushing a chessman on the board in violation of the rules of chess is a move in chess. In contrast, a violation of the strategic rules of a game may make you guilty of stupidity but not of an infraction.

Still other kinds of fallacies involve mistaking of one kind of interrogative game for another. For instance, in one kind of game a premium is put only on the truth and informativeness of the ultimate conclusion, while in another one the sole aim is to refute another persons' thesis. Many such distinctions can in our approach be implemented by different kinds of assignments of utilities to the different courses that an interrogative game can take, including their ultimate results. Then it may turn out to be a strategic mistake in a truth-seeking game to rely on the answers of any one oracle, even in cases where the oracle is your opponent. This mistake can be considered a form of the so-called fallacy of arguing *ad hominem*.

Yet another mistake involves a confusion between the different *dramatis personae* in an applied interrogative game. For instance, it is sometimes alleged that the *ad hominem* fallacy consists in "rejecting the claims a person advice simply on the basis of derogatory facts ... about the person making the claim." But if the person in question acts as an "oracle" in our technical sense, then it is of course according to Hoyle and not fallacious in the least to use derogatory facts about him to cast doubts on the testimony of such on oracle. In contrast, it is a mistake to try to evaluate someone's argument by reference to what one knows about the arguer.

We have a mistake here only if the role of an oracle is confused with the role of an inquirer. The mistake consists in confusing the evaluation of an oracle (e.g. a witness) with the evaluation of an arguer. An argument, unlike the testimony of witness, has to be evaluated *ad argumentum*, not *ad hominem*, to use the traditional locutions.

ACKNOWLEDGEMENT

This article represents independent work by the junior authors based on suggestions by the senior author. Ilpo Halonen has been exclusively responsible for section 19 and Arto Mutanen for section 9.

REFERENCES

Only works actually referred to are listed here.

Belnap, Nuel, and T. B. Steel, 1976, *The Logic of Questions and Answers*, Yale U. P., New Haven.

Beth, E. W., 1953, "On Padua's Method in the Theory of Definition," *Indagationes Mathematicae* vol. 15, pp. 330–339.

Beth, E. W., 1955, "Semantic Entailment and Formal Derivability," Mededelingen van de Koninklijke Nederlandske Akademie van Wetenschappen, Afd. Letterkunde, N. R. vol. 18, no. 13, pp. 309–342.

Bromberger, Sylvain, 1992, *On What We Know When We Don't Know: Explanation, Theory, Linguistics, and How Questions Shape Them*, University of Chicago Press.

Collingwood, R. G., 1933, *An Essay on Philosophical Method*, Clarendon Press, Oxford.

Craig, William, 1957(a), "Linear Reasoning: A New Form of the Herbrand-Gentzen Theorem," *Journal of Symbolic Logic* vol. 22, pp. 250–268.

Craig, William, 1957(b), "Three Uses of the Herbrand-Gentzen Theorem in Relating Model Theory and Proof Theory," *Journal of Symbolic Logic* vol. 22, pp. 269-285.

Harrah, David, 1984, "The Logic of Questions," in D. Gabbay and F. Guenther, editors,

Handbook of Philosophical Logic vol. 2, D. Reidel, Dordrecht, pp. 715–764.

Hempel, Carl G., and Paul Oppenheim, 1948, "Studies in the Logic of Explanation," *Philosophy of Science* vol. 15, pp. 135-175.

Hintikka, Jaakko, 1976, *The Semantics of Questions and the Questions of Semantics (Acta Philosophica Fennica* 28, no. 4), Societas Philosophica Fennica, Helsinki.

Hintikka, Jaakko, 1984, "The Logic of Science as Model–oriented Logic," in P. Asquith and P. Kitcher (eds.), *PSA 1984,* 1, Philosophy of Science Association, East Lansing, pp. 177–185.

Hintikka, Jaakko, 1985, "A Spectrum of Logics of Questioning," *Philosophica* vol. 35, pp. 135–150.

Hintikka, Jaakko, 1987, "The Fallacy of Fallacies," *Argumentation* vol. 1, pp. 221–238.

Hintikka, Jaakko, 1988, "What Is the Logic of Experimental Inquiry?," *Synthese* vol. 74, pp. 173-188.

Hintikka, Jaakko, 1989, "The Role of Logic in Argumentation," *The Monist* vol. 72, no. 1, pp. 3-24.

Hintikka, Jaakko, 1991, "Towards a General Theory of Identifiability," in James H. Fetzer et al., editors, *Definitions and Definability,* Kluwer Academic Publishers, Dordrecht, pp. 161–183.

Hintikka, Jaakko, 1992, "The Interrogative Model of Inquiry as a General Theory of Argumentation," *Communication and Cognition* vol. 25, nos. 2-3, pp. 221-242.

Hintikka, Jaakko, 1995, "The Games of Logic and the Games of Inquiry," *Dialectica* vol. 49, 229–249.

Hintikka, Jaakko, 1996(a), "On the Development of Aristotle's Ideas of Scientific Method and the Structure of Science," in William Wians, editor, *Aristotle's Philosophical Development: Problems and Prospects*, Rowman and Littlefield, Lanham, Maryland, pp. 83–104.

Hintikka, Jaakko, 1996(b), *The Principles of Mathematics Revisited*, Cambridge U. P.

Hintikka, Jaakko, and Ilpo Halonen, 1995, "Semantics and Pragmatics for Why-Questions," *Journal of Philosophy* vol. 92, pp. 636–657.

Hintikka, Jaakko, and Ilpo Halonen, forthcoming, "Interpolation as Explanation."

Hintikka, Jaakko and Stephen Harris, 1988, "On the Logic of Interrogative Inquiry," in A. Fine and J. Lepkin (eds.), *PSA 1988,* 2, Philosophy of Science Association, East Lansing, pp. 233-240.

Hsiao, Cheng, 1983, "Identification," in Z. Griliches and M. D. Intriligator, editors, *Handbook of Econometrics*, vol. 1, chapter 4, North-Holland, Amsterdam.

Wisniewski, Andrzej, 1995, *The Posing of Questions: Logical Foundations of Erotetic Inferences*, Kluwer Academic, Dordrecht.

Piatelli-Palmarini, Massimo, 1994, *Inevitable Illusions*, John Wiley and Sons, N. Y.

Pitt, Joseph C., editor, 1988, *Theories of Explanation*, Oxford University Press, New York.

Robinson, Abraham, 1963, *Introduction to Model Theory and the Metamathematics of Algebra*, North-Holland, Amsterdam.

Robinson, Richard, 1971, "Begging the Question 1971," *Analysis* vol. 31, pp. 113–117.

Salmon, Wesley C., 1990, *Four Decades of Scientific Explanation*, University of Minnesota Press, Minneapolis.

4

WHAT IS ABDUCTION? THE FUNDAMENTAL PROBLEM OF CONTEMPORARY EPISTEMOLOGY

It is sometimes said that the highest philosophical gift is to invent important new philosophical problems. If so, Peirce is a major star on the firmament of philosophy. By thrusting the notion of abduction to the forefront of philosophers' consciousness he created a problem which — I will argue — is the central one in contemporary epistemology.

Now what is the notion of abduction, what is new about it, and why is it a problem? Peirce's notion and some of the problems it raises have recently been summed up by Tomis Kapitan (1997, pp. 477-478) in four theses:

Inferential Thesis. Abduction is, or includes, an inferential process or processes (5.188-189, 7.202).

Thesis of Purpose. The purpose of "scientific" abduction is both (i) to generate new hypotheses and (ii) to select hypotheses for further examination (6.525); hence, a central aim of scientific abduction is to "recommend a course of action" (MS 637:5).

Comprehension Thesis. Scientific abduction includes *all* the operations whereby theories are engendered (5.590).

Autonomy Thesis. Abduction is, or embodies, reasoning that is distinct from, and irreducible to, either deduction or induction (5.146).

In his earlier work, Peirce identified abduction with Aristotelian *epagoge* and interpreted it as an inverse of a syllogism. According to Kapitan (1997, p. 480),

> Peirce's 1878 model of reasoning by "hypothesis" from result and rule to case (2.623) is familiar:
>
> (F1) All *As* which are *B* are *C*. (Rule)
> This *A* is *C*. (Result)
> Therefore, the *A* is *B*. (Case)
>
> Abduction differs from deduction which moves from rule and case to result and from induction which goes from case and result to rule, and it is this tidy contrast in terms of syllogistic permutations that very likely led Peirce to trichotomize inference in the first place.

But in such a scheme an abductive inference does not necessarily yield even probabilistic support for its conclusion. Hence it is extremely puzzling why the early Peirce should have claimed that abduction in his sense is an inference. (Cf. the Inferential Thesis above.)

Moreover, the inverse syllogism model cannot be seriously claimed to be the only way in which new hypotheses and theories are engendered in science. (Cf. the Comprehension Thesis above.) This model is a special case of the idea, to be discussed below, that abduction is an inference to the best explanation. Peirce's early schema of abduction presupposes, if he is to claim universality for it, that all the best explanations are syllogistic. We know better than that, and Peirce came to know better, too.

But his later sense of abduction does not seem to fare much better. For one thing, it still is extremely difficult to see why abduction, whatever it is or may be, can be not only a rational operation but even a logical inference, in any sense of logical inference.

This puzzle can be said to be created by the combination of the different theses which Peirce defended and which were listed above. Abduction is ampliative, according to the Theses of Purpose and Comprehension. Hence it cannot be deductive, for valid deduction is tautological in the sense of not yielding any new information, such as new hypotheses. Hence abduction is not necessarily truth-preserving. Furthermore, it is contrasted by Peirce to induction, which according to him is the process of testing the new hypotheses provided by abduction. For this reason, abduction cannot very well provide even probabilistic support for its output. In other (more general) words, all the factors that might affect the reliability of abductive hypothesis formation belong to the inductive phase of inquiry not to the abductive one. In what sense can abduction then be an inference?

This puzzle about what abduction really is is deepened by Peirce's explicit acknowledgment of the presence of a conjectural element in abduction. It even seems that our abductive hypothesis-forming power is nothing but a mysterious power of guessing right. Indeed, Peirce himself writes as follows:

> In very many questions, the situation before us is this: We shall do better to abandon the whole attempt to learn the truth, however urgent may be our need of ascertaining it, unless we can trust to the human mind's having such a power of guessing right that before very many hypotheses shall have been tried, intelligent guessing may be expected to lead us to the one which will support all tests, leaving the vast majority of possible hypotheses unexamined. Of course, it will be understood that in the testing process itself there need be no such assumption to mysterious guessing-powers. It is only in selecting the hypothesis to be tested that we are to be guided by that assumption.

But how can an exercise of such "mysterious guessing-powers" possibly be an inference? Elsewhere Peirce speaks of a *method* of making good conjectures instead of a power of doing so. Sometimes he also uses the word *habit* in this context. I will return to these formulations later.

The conjectural element in Peirce's notion of abduction is precisely the kind of reason that has led other philosophers of science to embrace a hypothetico-deductive model of science. (Needless to say, we might equally well speak here of a hypothetico-deductive model of knowledge acquisition in general.) For nothing seems to be less rational and less subject to rules than guessing. Thus Peirce's idea of abductive inference as the source of all new hypotheses stands in a stark contrast to any hypothetico-deductive theory of science.

Here you can begin to see what I have in mind by calling the problem of abduction the basic question of contemporary epistemology. The most general problem to which both the hypothetico-deductive approach and the idea of abduction are attempted solutions is: What is ampliative reasoning like? Purely logical (in the sense of deductive) reasoning is not ampliative. It does not give me any really new information. Yet all our science and indeed our whole life depends on ampliative reasoning. But what is that reasoning really like? When we speak of the reasonings of the likes of Sherlock Holmes or Nero Wolfe as "deductions" accomplished by means of "logic," we do not mean philosophers' deductive logic which is not ampliative. But what are they, then? The hypothetico-deductive approach tries to brush them all under the carpet of "contexts of discovery" which cannot be dealt with by means of logical, epistemological or other rational means. Such a way of thinking might be congenial to the cultists of the irrational, but it is deeply dissatisfying in that it leaves unexamined an important part of the cognitive lives of all of us and hence is deeply un-Socratic.

The identification of the problem of Peircean abduction with the problem of the nature of ampliative inference is largely justified by Peirce's Comprehension Thesis (see above).

Before essaying my own answer, it is in order to deal with a widespread alternative interpretation of abduction. According to this interpretation, abduction is an inference to the best explanation.

This idea has a great deal of initial plausibility. In fact, abductive inference is often, perhaps typically, related to explanation. Peirce already emphasized that the new hypothesis that abduction yields should explain the available data. (See e.g. "Pragmatism as the Logic of Abduction" (May 14, 1903), pp. 15-16 in Peirce's numbering; Kapitan 1997, pp. 480-481.) Others have later strengthened the role of explanation in the definition of abduction and identified abduction with what is known as "inference to the best explanation."

This view is seriously oversimplified at best. Part of the difficulty can be seen by asking, first, what explanation is or, perhaps more pertinently, what explaining is. Most people who speak of "inferences to best explanation" seem to imagine that they know what explanation is. In reality, the nature of explanation is scarcely any clearer than the nature of abduction. I have argued elsewhere (See Halonen and Hintikka, forthcoming) that explaining a certain explanandum E is to derive it from an assumed background theory T plus a number of contingent truths A which are relative to E and which have to be found in order for an attempt to explain E is to succeed. An explainer's job description is thus twofold: on the one hand to find the auxiliary facts A and on the other hand to deduce the explanandum from them together with the background theory T. This entire process, including both the search for A and the derivation of E from T and A, can be thought of as an interrogative inquiry in the sense that the contingent data A can be conceptualized as nature's answers to the inquirer's questions. Certain further conditions may have to be satisfied by T and A in relation to E.

The pertinent features of explanation thus include the fact that the theory T is not in any literal sense a generalization from the different explananda E_1, E_2, ... that it can help to explain, for in each case E_i is implied by T only in conjunction with the *ad hoc* data (nature's answers) A_i, which may be different for different values of i, i.e. for different explananda.

What follows from these observations for the interpretation of abduction as inference to the best explanation? The first pertinent question here is: Explanation of what facts? The merits of a theory or hypothesis include its ability to explain new, previously unknown facts. But these facts will be, if they are genuine new ones, unknown at the time of the abduction, and even more so must the auxiliary data which help to explain them be unknown. Hence these future, so far unknown explananda cannot be among the premises of an abductive inference. For Peirce makes it clear that according to his lights abduction, like any inference, leads to new knowledge on the basis of what is already known. (Cf. below.) Moreover, an inference is a conscious operation, wherefore this earlier knowledge must be explicit at the time of the inference. Hence the unknown explananda cannot be what "an inference to the best explanation" is calculated to explain, for they are not in the conscious control of the reasoner at the time of the inference.

This observation is worth elaborating. Quite often, scientists frequently become aware of interesting phenomena explainable by means of a theory of interesting phenomena only through the theory itself, once it has been discovered. This happens, for instance, when a theory predicts the existence of an entity, such as a certain kind of subatomic particle, which then is subsequently discovered. The theory admittedly explains the existence of the particle, but the theory can scarcely be said to have been arrived at by means of inference to such explanations. The same goes for the existence of new phenomena. For an amusing example, Einstein was aware of the possibility of Brownian motion and thinking about the laws governing it before he even knew that Dr. Brown actually had observed such a motion. (See e.g. Fölsing, 1997, pp. 128-131.) If Dr. Brown had not done so, Brownian motion would have been observed in connection with the testing of Einstein's explanation. In such cases, a scientist's reasoning looks less like an inference from certain explananda to a hypothesis that explains them than an inference from an abductive hypothesis to the existence a new phenomenon it turns out to explain.

Einstein's discoveries provide still further examples. Needless (perhaps) to say, the discovery of the general theory of relativity was not an inference to the best explanation of the perihelion movement of the planet Mercury and of the curvature of light rays in a gravitational field during a solar eclipse, even though they were the first two new phenomena explainable by the general theory.

Hence the explananda relevant to the idea of an "inference to the best explanation" must be data known to the scientist drawing an abductive inference at the time of the inference. Otherwise, the idea of abduction as an inference does not make much sense. But even when the explananda E and the background theory T are known to the scientist, he or she may not be aware of the explanation that T can yield of E, for instance because he or she is unaware of the auxiliary facts or is unaware of the derivation of E from T and A. For instance, in the discovery of the outer planets the conclusion of the crucial abductive inference is naturally taken to be the existence of a new planet. But the mere existence assumption does not explain the apparent irregularities in the motions of known planets which are the starting-point of the discoverer's line of thought. In order to reach the explanation, the orbit of the new planet has to be established and its influence on the known planets calculated. But these tasks were not accomplished when the initial existence assumption was proposed.

Hence, since the abductive reasoner does not always have at his or her disposal explanations even of the known data, the abductive inference cannot be a step to the known data to a hypothesis or theory that best explains them.

Moreover, many of the most important types of scientific reasoning cannot be described as inferences to the best explanation in the first place. For instance, when a controlled experiment produces a dependence law telling us how the observed variable depends on the controlled one the law does not *explain* the result of the experiment. It is the *result* of the experiment, nature's answer to the experimental investigator's question. Thus the claim that inference to the best explanation is the only or merely a typical way of forming new hypotheses in science is simply false. An experimental formation of new laws has an excellent claim of being the typical way of forming new laws in experimental sciences.

For another instance or bunch of instances, many of the scientific inferences that presumable should be thought of as abductive are not generalizations from particular cases or explanations of such particular cases. They may for instance be successful syntheses of two earlier laws or theories which might even have seemed to be irreconcilable. In such cases, the wider new theory unifying the earlier ones does not always "explain" the earlier ones in any natural sense of the word. It reconciles them with each other. (Cf. here Hintikka, 1993(a).)

An instructive example is offered by the special theory of relativity. The old-fashioned philosophical view would have sought to interpret Einstein's theory as an attempt to explain certain "anomalies" left unaccounted for by the earlier views. And an explanandum apparently serving this role was in fact available in the form of the famous Michelson-Morley experiment. Einstein's theory did in fact in the light of hindsight explain why a terrestrial observer appears to be stationary with respect to the ether. Alas, historical scholarship has revealed that the Michelson-Morley experiment played no role in Einstein's actual line of thought that led him to the special theory of relativity. (See Holton, 1969; Fölsing, 1997, pp. 217-219.)

The most instructive way of looking at Einstein's discovery is to see it as a way of reconciling Maxwell's electromagnetic theory with Newtonian mechanics. This is reflected in the very title of his famous paper "Zur Elektrodynamik bewegter Körper." But it would be ridiculous to say that Einstein's theory "explains" Maxwell's theory any more than it "explains" Newton's laws of motion.

Hence the first and crucial step to a scientific hypothesis or theory that abduction is supposed to be cannot be thought of as an inference to the best explanation.

Defenders of abduction interpreted as "inference to the best explanation" sometimes defend their idea by appealing to examples from the history of science. Thus John R. Josephson (Josephson & Josephson, 1994, pp. 7-8) writes:

> Abductions are common in scientific reasoning on large and small scales. The persuasiveness of Newton's theory of gravitation was enhanced by its ability to explain not only the motion of the planets, but also the occurrence of the tides.

Such appeals to history are superficial, and on a closer examination easily turn against the appealer. In the cases referred to by Josephson, Newton most definitively does not present his discoveries as results of abduction (or equivalent). Abduction is described by Peirce and by others as a hypothesis-forming operation. Newton blandly denies that he is "feigning hypotheses." At times Newton goes as far as to say that according to him laws are "deduced" from phenomena.

The interpretation of abduction as an inference to the best explanation is also in conflict with what Peirce says in so many words. In the passage quoted above from 6.525 he says that in abduction one hypothesis may be preferred over others if the preference is *not* based on previous knowledge. But the whole idea of inference to the best explanation is that the choice is determined by the facts that are to be explained — that the outcome is the best explanation of these particular data.

Some defenders of the idea of inference to the best explanation have even tried to subsume enumerative induction under such a procedure (e.g. Harman 1968). This was in effect Peirce's early perspective on abduction, but it was not Peirce's mature view. On the one hand for Peirce, induction is essentially a procedure of testing and confirming hypotheses arrived at by abduction.

On the other hand, abduction was for Peirce the only way of introducing new hypotheses into inquiry. (Cf. the Comprehension Thesis above.) Yet nobody has in his wildest dreams suggested that all our new general knowledge is first arrived at by enumerative induction.

Before venturing my own solution to the problem of abduction I will first call attention to a dimension of this problem that has not yet come up. This dimension is present in Peirce, albeit not fully articulated. One context in which it can be seen is Peirce's notion of inference. It has already been seen that it is not quite what we are used to. As Peirce himself puts it.

> I call all such inference by the peculiar name, *abduction*, because its legitimacy depends upon altogether different principles from those of other kinds of inference.

What are those "altogether different principles?" Peirce's views on inference have been summarized as follows (Kapitan, 1997, p. 479):

(1) Inference is a conscious, voluntary act over which the reasoner exercises control (5.109, 2.144).

(2) The aim of inference is to discover (acquire, attain) new knowledge from a consideration that which is already known (MS 628: 4).

(3) One who infers a conclusion *C* from premiss *P* accepts *C as a result* of both accepting *P* and approving a general *method* of reasoning according to which if any *P*-like proposition is true so is the correlated *C*-like proposition (7.536, 2.444, 5.130, 2.773, 4.53-55, 7.459, L232:56).

(4) An inference can be either *valid* or *invalid* depending on whether it follows a method of reasoning it professes to and that method is conductive to satisfying the aim of reasoning, namely, the acquisition of truth (2.153, 2.780, 7.444, MS 692: 5).

The most interesting aspect of Peirce's notion of inference is (4). Usually the validity and the other merits of an inference are judged in terms of the relation of the premise or premises to the conclusion, for instance whether the truth of the premises guarantees the truth of the conclusion or perhaps whether it makes the conclusion probable to a certain degree. The term "rule of inference" is usually restricted to cover only such inferences as can be justified in terms of the premise-conclusion relation either because the step from the premises to a conclusion is truth-preserving or because the premises make the conclusion probable to a certain degree. Peirce is making a much more important break with this traditional idea than he himself seems to realize. He is going beyond rules of inference that depend on the premise-conclusion relation alone and is considering also rules or principles of inference "of an altogether different kind." They are justified by the fact that they exemplify a method that is conducive to the acquisition of new knowledge.

Hence there are two different kinds of rules (principles) that can justify an inference. Peirce does not seem to distinguish these two kinds of rules or principles from each other systematically. It would have been most helpful, however, if he had done so. I have called the former kind of rules *definitory rules* and assimilated them to the rules that define a strategic game like chess — or like deduction or scientific inquiry, for that matter. Such definitory rules are merely permissive. They tell us what moves one may make in given circumstances, but they do not tell anything about which moves are good, bad or indifferent. Such advice is codified in what I have called *strategic rules* (or principles). From the general theory of games we know that such rules cannot normally be formulated in move-by-move terms, for instance in terms of the relationship of premises to a conclusion, only in terms of complete strategies. (A game theorist would express this point by saying that in general utilities can absolutely speaking be associated only with entire strategies, not with particular moves.)

Now we can see that Peirce's vantage point contained at one and the same time a brilliant insight and a serious limitation. The insight was into the importance of strategic rules. For what Peirce's statement quoted above amounts to is that the validity of an abductive inference is to be judged by strategic principles rather than by definitory (move-by-move) rules. This is what makes an abductive inference depend on its legitimacy "upon altogether different principles from those of other kinds of inference." What these "different principles" were in Peirce's mind can be gathered from his various statements. One typical expression of the difference is Peirce's distinction between the validity and the strength of an argument.

> … it is only in Deduction that there is no difference between a *valid* argument and a *strong* one. ("Pragmatism as a Logic of Abduction," p. 17.)

Thus an argument can be logical but weak. Such statements leave little doubt that the kind of validity Peirce had in mind was essentially strategic.

There nevertheless was a limitation that handicapped Peirce's thinking is his failure to appreciate fully the difference between definitory and strategic rules. Whenever he tries to explain the kind of validity that does not go together with strength he becomes hesitant and resorts to examples. He never gives a general characterization of the difference between what I have called definitory and strategic rules. These two kinds of rules do not deal with different kinds of inference. They are different kinds of rules governing the same kinds of inferential steps. Peirce recognizes the difference between the ways in which definitory and strategic rules are legitimized or validated. Definitory rules of inquiry are validated insofar as they confer truth or high probability to the conclusion of each particular application of theirs. In contrast, strategic rules of inquiry are justified by their propensity to lead the inquirer to new truths when consistently pursued as a general policy. It is worth emphasizing that this propensity to lead to new knowledge must not be assimilated to an inference rule's ability to confer truth or high probability to its conclusion in each particular case. The former propensity can for instance be manifested when a rule application opens up new future possibilities of knowledge acquisition even when it does not itself provide the inquirer with any new truths or even new hypotheses. Peirce seems to have realized fairly clearly what a strategic justification of a rule application consists in. Indeed, he seems to see the justification not only of abductive but also inductive steps as strategic.

> *Induction* is reasoning which professes to pursue such a method that, being persistent in, each special application of it ... must at least indefinitely approximate to the truth about the subject in hand, *in the long run.* [Emphasis added.] *Abduction* is reasoning, which professes to be such, that in the case there is any ascertainable truth concerning the matter in hand, the general method of this reasoning though not necessarily each special application of it must eventually approximate the truth. (Eisele, ed. *The New Elements* vol. 4, p. 37.)

What Peirce does not realize is that one and the same step of reasoning, including deductive reasoning, can normally be considered both as an application of a definitory rule and as an application of a strategic rule — in fact, several alternative strategic rules that differ in their consequences for other particular cases. It is admittedly the case that deduction differs from abduction in the kind of justification of particular steps. As Peirce puts it,

> *Deduction* is reasoning which proposes to pursue such a method that if the premisees are true the conclusion will in every case be true. (Loc. cit.)

But this does not obliterate the fact that we can in deduction, too, distinguish definitory and strategic rules from each other. Furthermore, if abduction is to obey formal laws like any other inference, as Peirce believes, it must likewise be subject to definitory (formal) rules. In brief, what Peirce does not realize is that there is a definitory vs. strategic distinction that cuts across his trichotomy deduction-abduction-induction.

The same shortcoming can also be described by saying that even though Peirce recognized the vital importance of strategic rules in inquiry, he did not possess the general concept of strategy in the abstract sense which was first spelled out clearly by von Neumann (1928, see also von Neumann and Morgenstern, 1944). He had to try to make other concepts to do the job of the notion of strategy. How he did that requires a separate investigation. I believe that such an investigation might yield interesting insights into Peirce's ways of thinking and into the concepts he employed. For instance, it seems to me that the concept of habit was one of the notions he used to serve some of the same purposes as the notion of strategy has been introduced by later thinkers to serve. This would among other things help to understand why Peirce's concept of habit differs from its customary namesake in being a conscious operation of the human mind. I suspect, in other words, that inside each Peircean habit there lurks (at least in the area of epistemology) a strategic rule trying to get out.

In the light of hindsight, Peirce's failure to command the general concept of strategy can in any case be seen to manifest itself in other parts of his thought. For instance, as Risto Hilpinen first pointed out, Peirce put forward completely explicitly what is known as the game-theoretical interpretation of quantifiers. This is no mean feat, for this "interpretation" is in my judgment the only one that does justice to the actual logic of these crucial logical notions. Yet Peirce never incorporated the game-theoretical idea into his logical theory or otherwise put it to major uses. Why? The answer is in my view that on the game-theoretical interpretation the truth of quantificational sentences can only be conceptualized in terms of the strategies that the verifier has at his/her/its disposal. These strategies have their technical incarnation in the form of what is known as Skolem functions. Alas, they were introduced only in the twenties, as was the general game-theoretical notion of strategy. (See here Hintikka, 1988.)

Another idea of Peirce's that can perhaps be interpreted as an indication of a need of some notion like strategy is his requirement that the aim of scientific abduction is to "recommend a course of action." For such recommendation can scarcely mean a preference for one particular action in one particular kind of situation, but presumably means a policy recommendation. Alas, again Peirce does not seem to have developed this idea further.

The distinction between definitory rules and strategic rules reveals an ambiguity in Peirce's characterization of the criteria of the validity of an inference. When is a method or rule "conductive to the acquisition of truth"? A definitory rule is when the truth of its premise or premises guarantees the truth (or at least the high probability) of its conclusion. But a strategic rule need not do so in order to further the aims of inquiry. It suffices for it to lead to truth in the long run. The answer to a strategically correct question might not provide any information that would by itself serve the ultimate end of the inquiry in question, and yet be instrumental in finding out the truth, for instance by providing the inquirer with the presupposition of a question which the inquirer could not have asked without it and which will directly further the cause of the inquiry.

Applied to Peirce, this seems to reveal an inconsistency in his position. On the one hand, he seems to say that all inferences have to be judged strategically, in that it is the propensity of the method they instantiate to lead to truth that is the criterion of their validity. On the other hand, he suggests that it is a peculiarity of abductive inferences that they are so judged, in that their legitimacy is judged by altogether different principles from other kinds of inference. For presumably those "altogether different" principles are the strategic ones.

Thus, recognizing that Peirce was considering abduction from an implicitly strategic vantage point does not solve all the problems concerning this notion. First and foremost, the distinction between definitory and strategic rules fails to help us to discover any separate class of inferences that might be called abductive. Any move in any strategic game can be considered both from a definitory and from a strategic vantage point. You can raise two different questions about it. First, was it according to Hoyle?[1] Second, did it make it easier for the player who made it to win? These questions can concern the very same move in the very same game, which can be a "game" of inquiry. The former concerns the status of the move from a definitory vantage point, the latter from a strategic one.

Peirce's unspoken emphasis on the strategic viewpoint helps to understand why he thought of abduction as a rational procedure and not as blind guessing. But it does not help us to identify a class of rules or even a class of steps in a line of thought that can be classified as abductive. Even less so does it help to explain why such steps could be assimilated by Peirce to inferences in the same sense as deductive inferences.

However, the idea of a strategic principle provides us with an Archimedean point by means of which we can identify the nature of abductive and in general ampliative inference. This Archimedean point is the requirement that any given ampliative step in a rational argument must be capable of being judged strategically. In an argument or line of rational thought, the information that a fresh step codifies either is contained in the earlier propositions in that line of reasoning, or else is (partly or wholly) new. In the former case, the step in question is a deductive one and should be studied in deductive logic. In the latter case, the step is ampliative. And if Peirce is right, it must be abductive, at least if the output is a significant new hypothesis.

What has to be known before such an ampliative step of reasoning can be rationally evaluated? This means: before the concept of strategy can be applied to it. What is it that a reasoner must be conscious of and have under his or her control in an abductive inference?

Recall that an ampliative step of reasoning brings new information to the argument. Hence the following must clearly be the case:

(1) It has to be known who or what was the source of the new information

Otherwise the reliability of the answer cannot be gauged, for this reliability obviously depends on the source.

(2) In order for the inquirer's action to be rational he or she must have intentionally chosen that particular oracle (source of information). As Peirce says, it has to be a

conscious, voluntary act. The inquirer's performance hence cannot be evaluated rationally unless it is known what other sources of information were available to the inquirer.

In other (more general) words, it has to be specified what the other moves are that were available to the inquirer.

(3) The inquirer cannot have known before the abductive step what the information is that the oracle provided him or her with, for otherwise the information would not have been new. However, the inquirer's move in turning to this particular oracle in this particular way would not have been fully in his or her conscious control and as a result could not be rationally evaluated unless the inquirer knew which other items of information could have resulted from his or her decision to consult this particular source of information.

(4) Furthermore, by the same token the inquirer must have known what results a consultation of each of the other different available oracles might have yielded.

But if (1)-(4) are satisfied, the new information might as well be thought of as an answer by the oracle to a question put to it (him, her) by the inquirer.

For instance, if the oracle consulted could have responded by the information specified by A_1 or A_2 or ... instead of A_o (the actual answer), then the inquirer's action might as well be construed as asking the question,

Is it the case that A_o or A_1 or A_2 or ··· ?

Thus all the new information flowing into a fully rational argument might as well be understood as so many answers to the inquirer's questions.

This, then, is my solution to the problem of abduction, which has meanwhile been generalized so as to become the problem concerning the nature of ampliative inference in general. Abductive "inferences" must be construed as answers to the inquirer's explicit or (usually) tacit question put to some definite source of answers (information).

This answer is not put forward as a total novelty. In spirit, and in some cases in letter, the picture of rational inquiry which embodies my answer is remarkably close to the questioning method (*elenchus*) of the Platonic Socrates. The same method constitutes (I have argued) the methodology of the early (pre-syllogistic) Aristotle. Much later, Collingwood (1944) and Gadamer (1975) have likewise recommended what they call the logic of question and answer as the crucial method of inquiry.

My solution to the abduction problem is not merely an abstract theoretical thesis. It has direct implications to the way epistemology and the general philosophy of science is to be studied. The key to them is the general theory of questions and answers. This theory can be developed in great detail and with a great deal of precision. For instance, the theory of scientific explanation will amount to a study of the logic of why-questions. And since the logical theory of questions and answers is obviously dependent on the logic of knowledge (epistemic logic), which should more aptly be called the logic of information, the basis of all epistemology should be epistemic logic, suitably developed.

This solution prompts a number of further observations. One of them is that it was in part anticipated by Peirce. In fact, Peirce puts forward himself the interpretation of abduction as an interrogative step.

> The first starting of a hypothesis and the entertaining of it, whether *as a simple interrogation* [emphasis added] or with any degree of confidence, is an inferential step which I propose to call *abduction*. (6.525)

Peirce continues:

> This will include a preference for any one hypothesis over others which would equally explain the facts, as long as this preference is not based upon any previous knowledge bearing upon the truth of the hypotheses, nor on any testing of any of the hypotheses, after having admitted them on probation. I call such inference by the peculiar name, *abduction*, because its legitimacy depends on altogether different principles from those of other kinds of inference.

Elsewhere, he writes:

> It is to be remarked that, in pure abduction, it can never by justifiable to accept the hypothesis otherwise than as an *interrogation*. [Emphasis added.]. But as long as that condition is observed, no positive falsity is to be feared. [Peirce, ed. by Buchler, p. 154.]

This passage is useful in that it refutes the interpretation offered by Isaac Levi of Peirce's notion of abduction as merely delineating a class of possible answers. Levi writes (1991, p. 71):

> The task of constructing potential answers to a question is the task of abduction in the sense of Peirce.
>
> The "conclusions" of abductions are conjectures that are potential answers to questions. Deduction elaborates on the implications of assumptions already taken for granted or of conjectures when they are taken, for the sake of argument, to be true.

> Induction weeds out for rejection some conjectures, leaving the survivors for further consideration. (Ibid. p. 77)

This is not quite the whole story, for Peirce's words in 6.525 (see above) show that for him in abduction one may prefer one possible conjecture over others. However, from a strategic viewpoint Levi may be right, in that the choice of the set of alternative answers amounts to the choice of the question to be asked. And it is that choice that is crucial strategically.

So why did Peirce not simply identify an abductive inference with a question-answer step in an interrogative inquiry? A fully confirmed answer to such a question is impossible to give. Nevertheless, Peirce's keen insight into the strategic aspects of inquiry, combined with his failure to separate the definitory and strategic aspects of inquiry from each other, suggests an instructive hypothesis. (Or perhaps I should rather say, suggests an instructive abductive question.) What can one say in general about the best choices of questions to be asked in inquiry? How can one choose between different possible questions that can be asked? It will be seen below that the choice is determined by the same principles as the choice of the optimal deductive inference in the same circumstances. This was enough for Peirce to assimilate abductive questions to inferences, especially as he maintained that inferences are in general validated by their strategic propensities. But in an obvious sense, inferences cannot be identified with question-answer steps *simpliciter*, for that would mean assimilating them to each other also definitorily. Hence it was natural — albeit incorrect in the literal definitory sense — for Peirce to consider abductions a special class of inferences. Of course, definitorily speaking there is a world of difference between raising one question rather than others and positing a conjectured answer to it. But if someone like Peirce focuses his attention to the strategic questions, he might very well look away from these definitory differences.

Sometimes Peirce nevertheless relates interrogation to the "inductive" testing of hypotheses rather than to the formation of hypotheses; see e.g. Eisele, editor, *Historical Perspectives* vol. II, p. 899.

This is *prima facie* different from what I have suggested. Yet it may not be impossible to reconcile with what I have argued. What I insist on that the abductive part of inquiry can be conceptualized as inquiry in the sense of interrogation I do not maintain the converse. Hence I am not excluding that what Peirce calls the inductive component of inquiry also involves interrogation. On the contrary, it fits in very well with the idea that interrogative inquiry, like Peircean inquiry, is a self-correcting operation.

The systematic implications of my answer to the question that figures in the title of this paper have been codified in what I have called the interrogative model of inquiry. The line of thought presented in this paper helps to show why this name is too modest in that all ampliative reasoning can be thought of as involving a question-answer sequence.

Among other things, the nature of explanations can be studied by considering them as answers to why-questions.

I will not try to expound this interrogative approach to epistemology here. A survey of some of the basic results is presented in Hintikka, Halonen and Mutanen (forthcoming). Instead, I will use the results of that theory to show how my answer to the abduction problems helps to put into a perspective the different things that Peirce says about abduction. Thus Peirce's views serve as a useful framework by reference to which I can explain some of the main features of the interrogative approach to inquiry.

Among the features of Peirce's views that can be understood and largely justified in terms of this framework are there are the following:

(i) Abduction is an inference or, as I prefer to put it, is like an inference.
(ii) Abduction is the only way of introducing new hypotheses or, as I prefer to put it, new information into an argument.
(iii) Abductive steps in an inquiry have to be judged on the basis of strategic principles rather than definitory ones.
(iv) Abduction is different from induction. Induction is not a form of ampliative inference (introduction of new hypotheses), let alone the only form.
(v) We can now understand also the conjectural element in Peircean abduction.

I can here discuss these points only selectively. In its general features, the interrogative logic of epistemology ("epistemo-logic') is strongly reminiscent of Peirce's ideas. Consider, as an example, item (iv). In the interrogative model there are in it three different kinds of steps, viz. deductive and interrogative steps plus self-critical steps in which one of the earlier answers (or initial premises) is (at least tentatively) rejected ("bracketed"). Naturally, allowing such bracketing steps must be accompanied by rules that authorize their mirror images, unbracketing steps. This is strongly reminiscent of Peirce's trichotomy of deduction, abduction and induction. In particular, we can from this vantage point see why Peirce describes the inductive stage of inquiry as involving the testing of hypotheses rather than a series of inductive inferences. In the interrogative model, the critical stage consists in bracketing and unbracketing earlier answers, which are precisely the outcomes (Peirce's "hypotheses") of what on my reinterpretation are abductive steps of inquiry.

The most interesting one of the five points listed above is the first one.

(i) As far as the status of abduction as a special form of inference is concerned, the basic conclusion here is: no, there is no such form of inference (in any natural sense of inference) as abduction. Abduction should be conceptualized as a question-answer step, not as an inference in any literal sense of the word. Peirce was entirely right in separating abduction both from deduction and from induction.

But this is not the end of the story. We are still left with an intriguing mystery to solve. If we approach this puzzle from the vantage point of Peirce interpretation, it can be formulated by asking: If abduction is so radically different from both of the usually recognized modes of inference, deduction and induction, why on earth did Peirce call it inference? Systematically speaking, essentially the same question can be put by asking: If abduction is an inference, there must be rules for such inferences. What are these rules like? Peirce himself acknowledges that all inference takes place according to a certain method. (Cf. above.) What is this method? It was seen that according to Peirce an abductive method must be justified by strategic principles. What principles?

Many philosophers would probably bracket abductive inference with inductive inference. Some would even think of all ampliative inference as being at bottom inductive. In this matter, however, Peirce is hundred percent right in denying the role of naked induction in forming new hypotheses. All the sharpest analysts from Hume on have pointed out that the justification of inductive arguments depends on the regularity of the courses of events in our universe. Such regularity assumptions are factual ones and can in principle be disproved by experience. Hence they can be known only *a posteriori*. And if they are spelled out and made explicit, we do not need any longer any specifically inductive rules of inference. Deductive ones, including the deductive inferences of probability calculus, do the trick perfectly well.

I have argued that the same conclusion can be drawn, appearances notwithstanding, from Carnap's heroic and in the last analysis self-defeating attempt to develop an *a priori* logic of induction. (Cf. Hintikka 1993(b).) What is perhaps more timely to emphasize in this day and age, various currently fashionable modes of ampliative reasoning are in the same boat with inductive inference, including sundry nonmonotonic and circumscriptive inferences. Accordingly, no modes of ampliative reasoning other than interrogative moves are included in our interrogative model of knowledge acquisition, just as in Peirce's trichotomy.

Hence the only remaining similarity that could even mistakenly motivate Peirce's calling abduction an inference must be between abduction and deduction. But my interrogative interpretation of abduction seems to destroy totally any such assimilation. For what could be more dissimilar than a deduction, which merely reshuffles and spells out previous information, and an interrogative step, in which the answer to a question brings in fresh information?[2]

Yet this dissimilarity is only skin deep or, more accurately speaking, only definition deep. Here the logic of questions and answers mentioned above performs a major service for epistemological analysis (and synthesis). One thing that it brings out is the need of recognizing the role of presuppositions in interrogative inquiry. Before the inquirer is in a position to ask, say, a *who*-question ("Who did it?"), he or she must establish its presupposition. ("Someone did it.") Hence on paper, as a transition from one proposition to another, an interrogative step looks rather similar to a deductive step. The latter takes the inquirer from a premise or premises to a conclusion, while the former takes the inquirer from the presupposition of a question to its answer.

Moreover, the very same sentence can serve as the presupposition of a question and as the premise of a deductive step. For instance, an existential sentence of the form

(1) $(\exists x)S[x]$

can either serve as the presupposition of the question

(2) What (who, when, where, …), say x, is such that $S[x]$?

or as the premise of an existential instantiation which introduces a "John Doe" like "dummy name" of an "arbitrary individual," say β. In the former case, the output of the relevant step is a sentence of the form

(3) $S[b]$

where b is a singular term, for instance a proper name. In the latter case, the output is of the form

(4) $S[\beta]$

Here (4) differs from (3) only by having a dummy name where in (3) there was a real name.

This analogy has not caught the eye of theorists yet, for the very good reason that it has not in the past been extended to all types of questions. It can be so extended, however, as I have shown in my largely unpublished work. (The only qualification needed here is that *why-* and *how*-questions have a different logic and therefore have to be dealt with separately; see Hintikka and Halonen, 1995.)

It seems to me that Peirce had an intuitive understanding of this type of similarity between abductive and deductive inferences. One of his main reasons for calling abductive steps "inferences" is that they have a "perfectly definite logical form," to quote "Pragmatism as a Logic of Abduction" (p. 15). Because of this definite logical form, an abductive inference cannot yield "quite new conceptions" (op. cit. P. 16), even though it (and it alone) can introduce new "hypotheses."

These similarities between questions (abductive steps) and logical inferences (deductive steps) are purely formal, however. An epistemological assimilation of the two to each other on the mere basis of such formal similarities would be irresponsible.

Behind these formal similarities there nevertheless lies a remarkable strategic similarity. How do questions of strategy enter into the concrete situations of an interrogative or deductive reasoner? In either case, at any one given stage of inquiry the reasoner has a number of propositions (sentences) available to be used as presuppositions or as premises. In either case, the proximate strategic question is: Which one to use first? If the inquirer is reasoning deductively, the crucial strategic first question is: Which sentence or sentences should I use as the premise or as the premises of a deductive inference? It can be shown that the most sensitive strategic question in deduction is: Which sentence should I use first as the premise of an existential instantiation or its generalization, functional instantiation? If the inquirer is reasoning empirically (interrogatively), the next strategic question is: Which one of the available sentences should I use as the presupposition of a wh-question? These candidate sentences are the very same ones that could be used as premises of existential instantiations, suitably generalized.

Neither question admits in general a mechanical answer, in the sense that there is in neither case any recursive function that always specifies an optimal choice. However, insofar as we are dealing with pure discovery in the sense that all answers taken to be reliable, there is a remarkable connection between the two choices. Even if there is no mechanical way of making the optimal choice in either case, it can be shown that *the best choice is the same in both cases*. The strategically best question that can be asked has as its presupposition the optimal premise of an existential instantiation in the same circumstances. In this sense, the strategic principles of abductive reasoning, interpreted as I have done, are the same as the strategic principles governing deduction.

This result not only explains why somebody like Peirce should call abductive steps of reasoning inferences; it vindicates such a practice (aside from purely terminological questions). Even though interrogative and deductive steps are definitorily different in a radical way, they are strategically speaking governed by the same principles, at least insofar as we can look away from the critical and evaluative aspect of inquiry that Peirce called inductive.

Thus the interrogative interpretation of abduction or more generally speaking the construal of all ampliative reasoning as interrogative, while in a sense vindicating Peirce's tripartite separation of reasoning into deductive, abductive and inductive inferences, at the same time but in a different sense reveals fundamental connections between the three.

These connections can be spelled out further in a different direction. It might seem that the critical and evaluative aspect of inquiry that Peirce called inductive still remains essentially different from the deductive and abductive aspects. A common way of thinking equates all ampliative inferences as inductive ones. Peirce was right in challenging this dichotomy. Rightly understood, the ampliative vs. nonampliative contrast becomes a distinction between interrogative (ampliative) and deductive steps of argument. As in Peirce, we also need over and above these two also the kind of reasoning that is involved in testing the propositions obtained as answers to questions. I do not think that it is instructive to call such reasoning inductive, but this is a merely terminological matter.

From the vantage point of the interrogative approach, Peirce's terminology can be claimed merely to follow ordinary usage when he calls an interrogatively interpreted abductive step an inference. The reasoning of the likes of Sherlock Holmes or Nero Wolfe are not deductive, nor does it conform to any known forms of "inductive inference." The "deductions" of great detectives are in fact best thought of as question-answer sequences interspersed by deductive inferences (I have argued). Yet people routinely call them "deductions" or "inferences" accomplished by means of "logic" and "analysis." They now turn out to be right strategically speaking, though not literally (definitorily) speaking. From the strategic vantage point we can say that thus any seriously asked question involves a tacit conjecture or guess.

We can also put into perspective Peirce's idea that there is a conjectural element in abductive inference. (Cf. (v) above.) From the strategic viewpoint, the crucial question about abductive questions is: Which one to ask first? So far I have given only a conditional answer, viz. that the choice of the best question is determined by the same principles as the choice of the optimal deductive inference in the same circumstances. But how are these coordinated choices made? Probably the only general answer, which unfortunately does not yield any directly applicable recipes, is to say that the choice of the right questions depends on one's ability to anticipate their answers. Strategically, there is little difference between selecting a question to ask in preference to others and guessing what its answer will be — and guessing how it compares with the expected answers to other questions that could be asked. And even in the case of deductive rules, the secret of a good strategist is to be able to anticipate where the inferences lead. This is especially clear when one is using what as known as *tableau* methods. There an attempted derivation of G from F is construed as an attempt to construct a model in which F is true but G false. One's success in doing so typically lies in being able to steer the construction to an obviously impossible configuration. And this ability to anticipate where the successive construction steps would lead the logician is the crux in his or her strategy selection.

Likewise, in interrogative inquiry, the crucial consideration is to anticipate the epistemic situation brought about by the answer — which in practice typically amounts to anticipating the answer. And such anticipation is hard to characterize in any terms other than guessing.

What is especially interesting — and especially reminiscent of Peirce — here is that such an element of guessing in abductive questioning is completely compatible with the strategic analogy between deduction and interrogation that inspired Peirce to call abduction an inference.

This interplay of deduction, questioning and conjecture has not been discussed very much by epistemologists, but it is an old hat for puzzle fans. If you open a typical book of what are known as "lateral thinking puzzles," you will find, besides an initial list of puzzles and a concluding section containing their solutions, an intermediate third section telling what questions you should ask to solve the problem. Of course, the questions do not necessitate the right answers, which means that the intended answer is in a sense conjectural. But if the puzzle is a good one, the natural answer is obvious once the right question has been asked. Here is an example (Sloane and MacHale, 1996, p. 32):

> A man walked into a bar and asked the barman for a glass of water. They had never met before. The barman pulled a gun from under the counter and pointed it at the man. The man said "Thank you" and walked out. Why should that be so?

One of the questions that could be asked here is: What was it that the man needed and that could be satisfied as well by a glass of water and by having a gun pointed at him? (If you still don't get the solution, look for help in the endnotes.)[2]

One more apparent contradiction in what I have said deserves a comment. Acknowledging the element of guessing in abduction does not gain say its character as inference subject to precise (strategic) rules. This would not be the first case where an apparently purely psychological phenomenon turns out to have an objective, rule-governed rationale. The first striking explanation of this kind was the von Neumann explanation of the apparently irrational trick of bluffing in games like poker as being nothing more and nothing less than the use of a mixed strategy. Here, too, guessing the right answer to different questions is little more than the phenomenological side of the same coin as the choice (or creation) of an optimal strategy.

These examples should be enough to convince the reader that my interrogative construal of Peircean abduction vindicates strikingly well what he says of abduction.

There is one more aspect of Peirce's views (and statements) that can serve to highlight the epistemological situation. It is the notion of hypothesis. Abduction is characterized by Peirce by the universal process of forming new hypotheses. I might seem to have departed from Peirce's intentions when I presented abduction as the universal method of introducing new information into a rational argument. The answer is that by information I do not mean necessarily true information. On the contrary, its being *new* information implies that it is not implied by what is already known (or at least accepted). In this sense, abduction as here construed always has a hypothetical element.

But there is even more than that to the hypothetical character of the conclusion of an abductive inference. This character, aptly emphasized by Peirce, illustrates one of the most important things about the justification of abductive inferences. According to Peirce, this justification lies in the abductive rule's being conductive to the acquisition of truth. Many contemporary philosophers will assimilate this kind of justification to what is called a reliabilist one. Such reliabilist views are said to go back to Frank Ramsey who said that a belief is knowledge if it is true certain and *obtained by a reliable process.*

Unfortunately for the reliabilists, such characterizations are subject to the ambiguity which was pointed out above. By a reliable process, one can mean either a process in which each step is conductive to acquiring and/or maintaining truth or closeness to truth, or one which as a whole is apt to lead the inquirer to truth. Unfortunately, most reliabilists unerringly choose the wrong interpretation, namely the first one. As was pointed out above, the true justification of a rule of abductive inference is a strategic one. And such a strategic justification does not provide a warrant for any one particular step in the process. Such a particular step may not in any obvious way aid and abet the overall aim of the inquiry. For instance, such a step might provide neither any new information relevant to the aim of the inquiry nor any new confirmation for what has already been established and yet serve crucially the inquiry, for instance by opening up the possibility of a question whose answer does so.

Furthermore, unfortunately for the reliabilists, the idea of a nonstrategic justification that they choose is not only mistaken but in the last analysis incoherent. From the theory of strategic processes misleadingly labeled game theory, it is known that what can be valuated (assigned "utilities" to) are in principle only strategies, not particular moves. Hence a theory of epistemic processes which operates with "warrants" for particular belief changes or other things that can be said of particular moves in our "games" of inquiry is inevitably going to be unsatisfactory in the long run.

One of the many things that Peirce's use of the term "hypothesis" can serve to highlight is precisely the strategic character of any justification of abduction. Being strategic, such justification does not *per se* lend any reliability to the outcome of some particular abductive inference. It leaves this outcome to the status of a hypothesis. Whatever reliability it may possess has to be established by the inductive component of inquiry.

Once again, it remains to be established precisely how clear Peirce himself was about this matter. I believe that he saw the point clearly enough, but was prevented from speaking out because he did not have an explicit notion of strategy at his disposal. This is why he did not fully explain what the "altogether different kind" of justification is that abductive rules can claim. In any case, Peirce disavows in so many words his own early identification of abduction as a species of probable inference. (See Kapitan, 1997, p. 493, note 1.)[3]

NOTES

[1] Translated from English into American, "Was it made in accordance with the definitory rules of the game?" (*Vide* e.g. Hubert Phillips, *The Penguin Hoyle, 1958.)*

[2] When was the last time you had a hiccup?

[3] In working on this paper, I have benefited from comments and suggestions by Risto Hilpinen.

REFERENCES

Peirce is referred to in the usual editions (*Collected Papers, Writings,* Eisele, Buchler, etc.).

I have also used the following Peirce MSS, copies of which were kindly made available to me by the Peirce Edition Project:

"On the Logic of drawing History from Ancient Documents, especially from Testimonies" (MS 690). This is a thick typescript that Peirce first handwrote in October and November 1901; it was typed by the secretary of his friend Francis Lathrop in early December 1901.

The seventh Harvard lecture of 1903, left untitled by Peirce (MS 315). In *Essential Peirce 2*, it will be titled "Pragmatism as the Logic of Abduction."

"An Essay toward improving our Reasoning in Security & in Uberty" (MS 682). This is one of several essays Peirce wrote at the end of his life on the topic of the fruitfulness of arguments.

Eight pages (numbered 4 to 11 by Peirce — pp. 1-3 are missing) from MS 683, an untitled manuscript that might have borne the same title as the previous document. It, too, bears upon the topic of uberty, and should also be dated September – October 1913.

Collingwood, R.G., 1946, *The Idea of History,* Oxford U.P.

Fölsing, Albrecht, 1997, *Albert Einstein: A Biography*, Viking, New York.

Freeman, Eugene, editor, 1983, *The Relevance of Charles Peirce*, The Hegeler Institute, La Salle, Illinois.

Gadamer, Hans Georg, 1975, *Truth and Method*, Continuum Publishing, New York.

Halonen, Ilpo, and Jaakko Hintikka, forthcoming, "Toward a theory of the process of explanation."

Harman, G., 1965, "Inference to the best explanation," *Philosophical Review* vol. 74, pp. 88-95.

Harman, G., 1968, "Enumerative induction as inference to the best explanation," *Journal of Philosophy* vol. 65, pp. 529-533.

Hilpinen, Risto, 1983, "On C.S. Peirce's theory of the proposition," in Freeman, editor (1983), pp. 264-270.

Hintikka, Jaakko, 1988, "On the development of model-theoretical viewpoint in logical theory," *Synthese* vol. 77, pp. 1-36.

Hintikka, Jaakko, 1993(a), "The concept of induction in the light of the interrogative approach to inquiry," in John Earman, editor, *Inference, Explanation and Other Frustrations,* U. of California Press, Berkeley and Los Angeles, pp. 23-43.

Hintikka, Jaakko, 1993(b), "On proper (popper?) and improper uses of information in epistemology," *Theoria* vol. 59, pp. 158-165.

Hintikka, Jaakko, and Ilpo Halonen, 1995, "Semantics and pragmatics for why-questions," *Journal of Philosophy* vol. 92, pp. 636-657.

Hintikka, Jaakko, Ilpo Halonen and Arto Mutanen, forthcoming, "Interrogative logic as a general theory of reasoning," in R. Johnson and J. Woods, editors, *Handbook of Applied Logic*, Kluwer Academic, Dordrecht.

Holton, Gerald, 1969, "Einstein, Michelson and the 'crucial' Experiment," *Isis* vol. 60, pp. 133-197.

Houser, Nathan, Don D. Roberts and James Van Evra, 1997, *Studies in the Logic of Charles Sanders Peirce,* Indiana U.P., Bloomington and Indianapolis.

Josephson, John R. and Susan G. Josephson, editors, 1994, *Abductive Inference: Computation, Philosophy, Technology*, Cambridge U.P.

Kapitan, Tomis, 1997, "Peirce and the structure of abductive inference," in Hauser et al, (1997), pp. 477-496.

Levi, Isaac, 1991, *The Fixation of Belief and Its Undoing*, Cambridge U.P.

Sloane, Paul, and Des MacHale, 1996, *The Lateral Logician*, Quality Paperback Book Club.

Tursman, Richard, 1987, *Peirce's Theory of Scientific Discovery*, Indiana U.P., Bloomington and Indianapolis.

von Neumann, John, 1928, "Zur Theorie der Gesellschaftsspiele," *Mathematische Annalen* vol. 100, pp. 295-320.

von Neumann, John, and O. Morgenstern, 1944, *Theory of Games and Economic Behavior*, Princeton U.P.

5

TRUE AND FALSE LOGICS OF SCIENTIFIC DISCOVERY

1. Introduction

This paper is a first step in a larger enterprise. The ultimate aim of my enterprise is to uncover the logical structures, in a strict sense of the word "logic", typically involved in scientific enterprise, not just in the justification of already obtained results but in the acquisition of new information.

Needless to say, such an enterprise is too large to be carried out in one paper. For those reason I shall in this paper restrict myself to a couple of partly preliminary matters. They include a discussion of certain competing views on the logic of scientific discovery and an acknowledgement of the most important existing treatments which partly anticipate the leading idea of my own approach.

In the title of this paper and in the paper itself, it is important to keep in mind that the word "logic" is to be taken quite seriously. The structures I am concerned with uncovering are intended to be of the same kind as the structures studied in mathematical logic and foundations of mathematics. I have a great deal of sympathy with the intentions of those philosphers who speak of "informal logic", but I don't think any clarity is gained by using the term "logic" for what they are doing.

2. False logics of discovery

But hasn't the logic of discovery already been discovered? Indeed, speaking of the true logic of scientific discovery presupposes that there are false ones on the market. This is in fact the case. Indeed, misconceptions concerning the subject are many and ripe. For instance, there are those who declare any logic of scientific discovery impossible, on the ground that there are no mechanical rules for the "lucky guesses" that are required for the advance of science. Their premise may be right, but the conclusion does not follow For by the same token there could not be any logic of deductive inference either, for it is well known that there are no mechanical procedures for finding desired logical proofs outside the most elementary department of logic.

In fact, if the philosophers who believe that there are no mechanical rules for scientific discovery want to prove their thesis instead of holding it as an unexamined dogma, they very likely will have to resort to some explicit formulation of the logic of inquiry themselves. Ironically, it is the logicians of science – that is, the philosophers who use explicit logical tools in their theories of the scientific procedure – who are in the best position to answer the question whether scientific discovery can be subjected to rules, to answer it negatively, and to actually *prove* their answer.

Some philosophers base the same impossibility claim on Thomas Kuhn's theory of scientific revolutions which is supposed to show that a major scientific breakthrough is not subject to logic because it means a shift in the "paradigm" which guides an entire research tradition and is not a subject of rational choice (1). But an appeal to Kuhn's actual views doesn't help them at all. In this matter, Kuhn is not a Kuhnian. Not only has he renounced the term "paradigm"; he is on the record as saying that the first people to take his theoretical ideas seriously were those philosophers, especially Sneed and Stegmüller (2), who have built formal models of theory change by means of set-theoretical conceptualizations (3). Indeed, the rise of this so-called structuralist school was the most important development in the philosophy of science in the seventies.

Unfortunately, the conception of the logic of science represented by the structuralist school also leaves a great deal to be desired. The main novelty of the structuralist approach is usually explained by saying that it lies in the replacement of the *statement view* of theories by a *structuralist view*. According to the former, so the story goes, theories are conceived of as sets of statements, whereas in the latter theories are characterized by their models.

This characterization is nevertheless misleading to the point of being perverse. It reduces the difference between the two views very nearly to a purely conventional choice of a jargon. On the one hand, in any view a set of statements specifies a set of models. This determination can in fact vary a great deal. For instance, a scientist might very well choose to accept only certain special models as the intended realizations of his assumptions. Moreover, and most importantly, the properties of statements studied in formal logic include those that serve these statements to determine (partly) their models (including certain aspects of the structure of those models). Even as simple a concept as logical implication has a neat model-theoretical import : X implies Y if and only if all the models of X are among those of Y. Hence the "statement view" is in reality perfectly compatible with the idea that theories are to be characterized in terms of their models.

If there is a real weakness in the so-called statement view, it is not that it operated with statements, but that the properties of statements it operated with were few and far apart and insufficiently grounded on model-theoretic considerations. One of those restrictions was the identification of a scientific theory with a deductive system. In brief, the logic of science of the statement view was wrong, not because of the role of statements, but because the logic this view presupposed was too poor.

The real question is therefore not whether the statement view should have been rejected but what the richer logic is that it should be replaced by. Some of the representatives of the so-called structuralist view answer this question in effect by going to the other extreme and using the full force of unrestricted set theory – to all practical purposes, the full force of classical mathematics – in theory conceptualizations. This is a very questionable procedure. It runs the risk of committing one of the same mistakes as the statement view committed, viz. severing the connection between, on the one hand, the actual formulation of a theory in some language or other, and the properties of its models on the other hand. For the model-theoretical properties of full set theory are extremely tricky, as is illustrated by the difficulty of various consistency and relative consistency problems. The "classical" structuralist view thus also fails to give a satisfactory answer to the question : What is the true logic of scientific theorizing".

Another way of looking at the predicament of the hard-line structuralists is the following : One important motive they have had is to keep their discussion independent of commitments to one particular language, and to concentrate simply and solely on the intended models of a theory. I find this idea very suspect. It is simply an illusion to think that we can deal with the models of a theory directly, without the help of some language or other. The real problem is not to dispense with references to a language, but to find the most suitable language to operate with one. The choice of this language is a delicate matter. It should be rich enough to do justice to actual scientific theorizing, and at the same time it should be simple enough so that its model-theoretic behavior can be mastered. It seems to me that even though the proper choice may be difficult, the resources of model theory and abstract logics have not been harnessed to the use of the philosophy of science in the way they would and should.

The illusion of language-independence is generated in the "structuralist" work by ample use of set theory without explicit reference to its axiomatization. This is not a satisfactory procedure philosophically, and I don't think that it is the best approach technically, either. Unrestricted (unaxiomatized) use of set theory is a difficult and untidy subject. Set theory is hence not the true logic of scientific inquiry.

Indeed, these viewpoints even suggest an answer to the question : if first-order languages are not rich enough, where should we look for richer ones ? The kind of answer my remarks bring close is to say : some extension of first-order languages which does not make the model theory of the outcome language hopelessly difficult. Candidates for this role include some of the simpler infinitary languages, for instance L_{w1w} in which one can for instance develop a neat model theory of probability theory (4). Other candidates which have not received sufficient attention include the outcome of imposing some further model-theoretical restraints on models of first-order sentences, e.g. restriction to saturated models or to prime models. Such restrictions add a great deal of expressive power, messing up the model theory of one's language.

An interesting fresh answer to this question has been put forward recently by Veikko Rantala and David Pearce (5). They propose to view scientific theorizing from the vantage point of recently developed theories of abstract logic and abstract model theory. This approach to the logic of science is very interesting and promising. We can expect that it will soon produce enough specific results to estimate its true worth.

I have a suspicion that there is ample room for further approaches, however. More specifically, I suspect that somewhat older and less fancy logical tools than those wielded by Rantala and Pearce can still give us plenty of mileage. It has become fashionable to criticize old-fashioned logic of science because it relied on first-order languages, and to criticize first-order languages as being too poor for the representation of actual scientific theories. This view may very well be true, but I doubt very much that it catches the most important parameter. I suspect that old-fashioned philosophy of science was relatively fruitless, not because the languages it dealt with were too poor, but became the model-theoretic and even proof-theoretic concepts it applied to these languages were too poor. This opens the door once again with vengance to the question : What are the richer concepts we should use ? What is the true logic of scientific discovery ?

3. Popper's non-logic of scientific discovery

In trying to answer this question, we don't get much help from Sir Karl Popper, either (6). According to his own principles, there is in the literal sense of the word no logic of scientific discovery. Not only is the title of his best known book a solecism according to his own principles, what is worse, Popper's constructive suggestions, whether we choose to pidgeonhole them under the heading "logic" or not, simply are not very helpful. In criticizing Carnap, Popper makes much of the inverse relation of probability and information. Since in the scientific enterprise we are seeking information (knowledge). Popper's advice for ascientist faced with the choice of a hypothesis or theory is : Be bold ! Choose the most improbable hypothesis !

Unfortunately, this is a seriously oversimplified picture. What is true is that the prior probability of a hypothesis and its absolute information are inversely related. However, hypotheses are never judged on the basis of their prior probability, but on the basis of their posterior probability (probability on evidence), while high information can very well mean high absolute one. A much better candidate for the role of an index of high informativeness is *expected* absolute information of a hypothesis. But if so, Popper's thesis is not generally valid. Given a suitable prior probability distribution, high posterior probability and high expected absolute information can very well go together (7). There simply does not exist the kind of inverse relation Popper believes in for the most interesting kinds of information and probability.

There exist mathematical results which show that boldness is one's best strategy if the odds are long against oneself (8). Such results may offer a way of vindicating partly some of Popper's ideas. This would make little difference to our overall evaluation of Sir Karl's non-logic of scientific discovery, however. It nevertheless enables me to formulate compactly what is wrong with Popper's philosophy of science. What the results I referred to mean is that it can be not only argued, but in a certain sense proved, that Popper's philosophy of science is, if not a counsel of despair, then a counsel of desperation.

4. Scientific procedure as information-seeking by questioning

Can we do better ? Yes, I believe that there were very good answers available to the question : What is the true logic of scientific discovery ? In order to find some of them, I start from an old idea which has never been exploited properly. The idea was put forward by Kant in the preface to the second edition of his *Critique of Pure Reason* (B xiii). In the scientific enterprise, Kant says there, reason "must not allow itself to be kept, as it were, in nature's leading-strings, but must itself show the way with principles of judgment based upon fixed laws, constraining nature to give answer to *questions* of reason's own determining." (Emphasis added.) In other words, the idea is to think of the scientific procedure as a series of questions the investigator puts to nature. By choosing these questions a scientist can direct the course of inquiry according to plans and principles of his of her own, like "an appointed judge who compels the witness to answer questions which he has himself formulated" (loc. cit.).

This idea of the scientific method as information-seeking by questioning has cropped up repeatedly in the literature. The best, and the best known, case in point is, undoubtedly Larry Laudan's book *Progress and its Problems* (9). Laudan rejects there all the traditional measures of the worth of theories, such as explanatory power, deductive power, high degree of confirmation, etc. In their place, Laudan puts a new criterion, the *problem-solving power* of a theory. As Laudan puts it himself in the first and foremost thesis of this book (p. 13),

> The first and essential acid test for any theory is whether it provides *acceptable answers to interesting questions* : whether, in other words, it provides satisfactory solutions to important problems. (Emphasis added.)

It seems to me that Laudan is here right in an important and fundamental way. However, I don't see that he explains the nature of this problem-solving or question-answering power as fully as one might hope. Laudan gives us a wealth of historical examples and explains carefully many qualifications which are important in applying the idea to actual historical material, such as the dependence of the class of relevant problems (questions) on pragmatic and historical factors. All this nevertheless leaves a philosophically demanding reader with a problem. What is this mysterious problem-solving power, anyway ? What are its structural characteristics, and how is it related to the logical structure of the theory in question? One way of looking at what I am trying to do here is to see my enterprise as an attempt to provide a real logic to spell out for the benefit of conceptually sophisticated philosophers what Laudan's ideas really amount to. In other words, I am trying to perform for Laudan the same service as Kuhn and Stegmüller sought to do for Thomas Kuhn, that is, to provide a real model for the suggestive but informal ideas put forward mostly in terms of historical examples.

In this spirit, then, my idea is to construe the scientific enterprise as a questioning procedure, as a series of questions put to nature or to some other source of information. There is much to be said about this model (or conceptualization) both on a general philosophical level and on a more detailed technical level. As my title indicates, I shall concentrate on one question only, which is : What kind of "logic" (in a fairly literal sense of logical structures involved) does the interrogative approach lead to ? What is on this model the real logic of scientific investigation ?

5. The logic of questions and answers

In a somewhat superficial sense, a partial answer is implicit in the basic idea of my entire approach. One kind of logic we clearly need here is the logic of questions and answers. Indeed, the reason why a questioning model of scientific inquiry has not been developed systematically before is that the crucial question-answer relation has remained almost completely unanalyzed. This has left in the air the precise issue which – as we in effect acknowledged above – is the nerve of the interrogative approach, viz. how a question conditions (partly determines) its answers. As the quote above shows, Kant was concerned with the way a researcher can by putting the right questions to nature guide her answers according to one's own plan.

In my theory of questions and answers, I offer an analysis of this crucial question-answer relation – both for conclusive (complete) answers and partial ones (10). Likewise, I have shown how the presuppositions of questions of different kinds are to be defined. This theory is here taken for granted. I don't have enough space here to expound it here. It is in any case only one ingredient in the logic of science on my interrogative model.

6. *Logic of questioning and semantical tableaux*

Another, still a somewhat superficial answer to the question as to what the logic of scientific investigation is, when it is construed as the logic of information-seeking by questioning, is obtained by considering what book-keeping methods one might use in one's questioning games against nature. The term "game" should be taken seriously here, for it turns out to be eminently useful to formulate our question-answer sequences in the form of games in the precise sense of the mathematical theory of games. One possible setup is the following : There are two "players" (parties) the scientist (also known as "myself") and nature. I start out with an initial assumption T, known as the "theory", and a conclusion C I am trying to prove. The premises I am allowed to use are then essentially T and the answers to the questions I put to nature one by one.

In such circumstances, the natural book-keeping method is to use a mild extension of Beth's semantical *tableaux* (11). (Beth's method is of course related extremely closely to Hintikka's model set method and to Gentzen's calculus of sequents.) (12) In the initial situation, T is put into the left column of the *tableau* and C in its right column. In making a move I have a choice between carrying out one step of *tableau* construction according to the usual rules and putting a question to nature and entering the answer in the left column of the *tableau* in question. (More explicitly, the answer is put in the left column of the *subtableau* in question, for a question is always asked by reference to one particular *subtableau.*)

Another variant of my information-seeking games against nature – and in some ways an even more natural one – differs from the one so far presented only in that my aim in the game is to prove either C or ~C. Then the semantical tableau used as our bookkeeping device is divided in two different ones, each having initially T in its left column, but the one having C in its right column while the other begins with ~C in its right column. Thus this interrogative game consists essentially in my attempt to answer the "big" initial question "Is it the case that C or ~C ?" by putting a number of "small" questions to my opponent.

Thus we have another relatively superficial answer to the question of what the logic of scientific information-seeking is on our interrogative model. It is an extension of the logic of Beth's semantical *tableaux.*

This answer is a step in the right direction. It enables us to formulate the explicit rules of the interrogative games of information-seeking, as we shall do below. It also enables us to show how actual examples of information-seeking by questioning can be "formalized", that is, represented in the canonical notation of semantical *tableaux.* (Of course, I am not "formalizing" them in the pejorative sense of turning them into uninterpreted calculi.)

7. The two roles of questions. Restraint on "small" questions

This does not yet close the question as to what the real logic of questioning and hence the real logic of scientific discovery is in the last analysis. The characterization given above of question-answer "games" is still incomplete in several crucial respects.

First and foremost, questions and answers enter into the scientific process in more than one way. Laudan is primarily emphasizing that the *aim* of a scientific theory is to solve important problems, to answer important questions. Kant's emphasis is on questions as a *means* of scientific inquiry. These two points are not incompatible, and are in fact combined in my *tableaux* formalization. When the aim of a questioning game is to prove C *or* ~C, we can say that the aim of the game is to answer a "big" question ("Is it the case that C or that ~C ?") by means of a number of "small" questions the investigator puts to nature.

The highlights the fact that certain restraints must be placed on the "small" questions (questions actually asked in the course of a game), lest the game be trivial. For if no restraints are imposed, a player can address at once to the source of information the question : "C or not ~C ?" and thereby trivialize the whole process. It has been shown by Richard Robinson (13) that Aristotle's injunction against *petitio principii* was, as the term suggests, intended precisely as a way of avoiding the trivialization of academic question-answer dialogues precisely in the way indicated.

In order to discover the real logic of the scientific process, we thus have to discuss the nature of the restraints that have to be imposed on the "small" questions to yes-or-no questions concerning the truth of atomic sentences (14).

8. Presuppositions of questions and questioning

Another important aspect of the logic of knowledge-seeking by questioning is the role of presuppositions of questions. It was already mentioned above in section 5 that my theory of questions and answers provides a definition of the presupposition of different kinds of questions. Again, it would take me too far to enter the details here. Instead, I shall call attention to a general distinction, which is at the same time a comment on earlier theorists of questions and answers.

Presuppositions of questions can enter into the tableaux formalization through the requirement that I can ask a question if its presupposition occurs in the left column of my *tableau.* This means that they have an important role in the determination of the optimal questioning strategies. One important aim of asking questions can be to get an answer which can serve as presupposition for promising further questions.

Over and above the entries to the left column of my *tableau* which come about in the course of a questioning game, there is in it also the initial theoretical assumption T (cf. section 6 above). Its logical consequences can serve as a presupposition to a suitable question, and in general it serves as a kind of presupposition (in a non-technical sense of the word) to the whole questioning enterprise.

This role of the original theoretical assumption T has not always been appreciated. For instance, in Collingwood's approach the role of the basic theoretical assumption T is assimilated completely to that of the presuppositions of particular "small" questions (15).

One of the crucial aspects of any satisfactory logic of the scientific enterprise, conceptualized as a question-answer sequence, will be an analysis of the precise role of the basic theoretical presupposition T of one's questioning procedure. One important fact is that some nontrivial theoretical assumption is often unavoidable. For instance, if the "small" questions are limited so as to pertain to the truth of atomic sentences only, only a limited number of "big" questions can be eventually answered. In contrast, it is known that, even if the same restriction is imposed on "small" questions, *every* "big" question (in a natural sense of the term) can be answered on the basis of a suitable "theory" T. even when the theory itself does not entail these answers (16). This illustrates vividly the role of the theoretical assumption T.

NOTES

(1) See T.S. Kuhn, *The Structure of Scientific Revolutions,* second ed., The University of Chicago Press, 1970; T.S. Kuhn, *The Essential Tension : Selected Studies in Scientific Tradition and Change,* The University of Chicago Press, 1977; Gary Cutting, editor, *Paradigms and Revolutions; Applications and Appraisals of Science,* University of Notre Dama Press,

Notre Dame, 1980; Ian Hacking, editor, *Scientific Revolutions* (Oxford Readings in Philosophy), Oxford U.P., 1981.

(2) See J.D. Sneed, *The Logical Structure of Mathematical Physics,* D. Reidel, Dordrecht, 1971; Wolfgang Stegmüller, *Theorienstruktur und Theoriendynamik,* Springer-Verlag, Berlin-Heidelberg-New York, 1973; Wolfgang Stegmüller, *Theory Construction, Structure, and Rationality,* Springer-Verlag, 1979. Useful surveys of the tradition Sneed and Stegmüller started are Ilkka Niiniluoto, The Growth of Theories : Comments on the Structuralist Approach in Jaakko Hintikka et al., editors, *Theory Change Ancient Axiomatics, and Galileo's Methodology,* D. Reidel, Dordrecht, 1980, pp. 3–47; and Ilkka Niiniluoto, "Scientific Progress", *Synthese* vol. 45 (1980), pp. 427–462.

(3) See *Erkenntnis* vol. 10, no 2 (July 1976), with contributions by Sneed, Stegmüller, and Kuhn.

(4) Cf. Dana Scott and Peter Krauss, "Assigning Probabilities to Logical Formulas" in Jaakko Hintikka and Patrick Suppes, editors, *Aspects of Indicative Logic,* North-Holland, Amsterdam, 1966, pp. 219–264.

(5) See David Pearce and Veikko Rantala, "On a New Approach to Metascience", *Reports from the Department of Philosophy,* University of Helsinki, no. 1 (1981), pp. 1–42 (with further references).

(6) Karl R. Popper, *The Logic of Scientific Discovery,* Hutchinson, London, 1959; Karl R. Popper, *Conjectures and Refutations,* Routledge and Kegan Paul, London 1963; Karl Popper, *Objective Knowledge,* Oxford U.P., 1972; P.A. Schilpp, editor, *The Philosophy of Karl Popper I–II,* Open Court, La Salle, Ill., 1974.

(7) See Jaakko Hintikka and Juhani Pictarinen, "Semantic Information and Inductive Logic" in Jaakko Hintikka and Patrick Suppes, editors, *Aspects of Inductive Logic,* North-Holland, Amsterdam, 1966, pp. 96–112; Jaakko Hintikka, "Varieties of Information and Scientific Explanation", in B. van Rootselaar and J.F. Staal, editors, *Logic, Methodology, and Philosophy of Science III : Proceedings of the 1967 Congress,* North-Holland, Amsterdam, 1968. Of course we have to restrict our attention to a relevant range of hypotheses, but there is nothing *ad hoc* about such restrictions.

(8) Cf., e.g., Lester E. Dubins and Leonard J. Savage, *How To Gamble If You Must : Inequalities for Stochastic Processes,* McGraw-Hill, New York, 1965.

(9) Larry Laudan, *Progress and Its Problems : Towards a Theory of Scientific Growth,* University of California Press, Berkeley, 1977; Larry Laudan, *Science and Hypothesis,* D. Reidel, Dordrecht, 1981; Larry Laudan, "A Problem-solving Approach to Scientific Progress" in Ian Hacking, editor (note 1 above).

(10) Jaakko Hintikka, *The Semantics of Questions and the Questions of Semantics* (Acta Philosophica Fennica, vol. 28, no. 4), Helsinki, 1976; "New Foundations for a Theory of Questions and Answers", forthcoming; "Questions with Outside Quantifiers" in Robinson Schneider et al., editors, *Papers from the Parasession on Nondeclaratives,* Chicago Linguistics Society, Chicago, 1982, pp. 83–92.

(11) E. W. Beth, "Semantic Entailment and Formal Derivability", *Mededelingen van de Koninklijke Nederlandse Akademie van Wetenschappen,* Afd. Letterkunde, N.R. vol. 18, no. 13, Amsterdam, 1955, pp. 309–342; reprinted in Jaakko Hintikka, editor, *Philosophy of Mathematics,* Oxford U.P., 1969, pp. 9–41.

(12) One can construe a *tableau* construction as an attempt to construct a model set in which the entries in the left *tableau* column are all included but which does not contain any entries in the right *tableau* column. If you turn a closed *tableau* upside down you obtain a Gentzen-type proof of the desired sequent.

(13) Richard Robinson, "Begging the Question 1971", *Analysis,* vol. 31 (1971), pp. 113–117.

(14) If my questioning games are thought of as research games against nature, the motivation of this restriction is clear. Nature can directly tell us what is true or false in particular cases, not whether some complicated sentence involving nested quantifiers is true or false.

(15) Cf. R. G. Collingwood, *An Essay on Metaphysics,* Clarendon Press, Oxford, 1940; R. G. Collingwood, *An Autobiography,* Clarendon Press, Oxford, 1939; Michael Krausz, "The Logic of Absolute Presuppositions" in Michael Krausz, editor, *Critical Essays on the Philosophy of R. G.*

Collingwood, Clarendon Press, Oxford, 1972, pp. 222–240.

(16) This is shown by the existence of theories which are model-complete but not complete. Cf., e.g. Abraham Robinson, *Introduction to Model Theory and to the Metamathematics of Algebra*, North-Holland, Amsterdam, 1963.

6

A SPECTRUM OF LOGICS OF QUESTIONING

1. Interrogative games as illustrating the general problems of dialogue theory

This paper* is prompted by my belief that, before we try to develop a general theory of dialogues, it is advisable to try to gain first some real insights into some of the special kinds of dialogues which the prospective general theory eventually will have to cover.[1] The case study carried out in this paper concerns one particularly simple kind of dialogue, viz. question-answer sequences where all the questions are addressed by the same speaker, called the Inquirer, to the same answerer, called Nature.[2] The only important additional element is that the Inquirer is also allowed to draw logical inferences from Nature's answers plus an initial theoretical premise T. These simple dialogues were first devised as models of information-seeking (empirical inquiry). Here they are studied also as special cases of a potential general theory of dialogues, illustrating the richness of the phenomena such a theory will have to cover.

2. Interrogative games defined.

The main rules and other specifications of the question-answer dialogues to be studied here can be formulated by considering them as two-person games between the Inquirer and Nature. The following is a list of some of the most important specifications :

Players: There are two players, called the Inquirer and Nature.

Scorekeeping: The game is conducted by reference to two semantical *tableaux* in E. W. Beth's sense.[3]

Initial position: In both *tableaux* there is a theoretical premise T in the left column. One of the two *tableaux* has C in its right column, the other one ~C.

The aim of the game: The Inquirer is trying to close one of the two *tableaux;* Nature is trying to prevent closure. In other words, the Inquirer is trying to answer the principal or initial question "C or not-C?".

Moves: At each stage, the Inquirer has a choice between two different kinds of moves, a deductive move and an interrogative move. Each move is relative to the stage which some one *subtableau* has reached, and adds a formula to it.

Deductive moves: A deductive move is a step of *tableau* building in accordance to the usual rules of *tableau* construction. It is assumed that they do not contain any rules which can violate the subformula principle, such as the cut rule, *modus ponens,* etc.[4]

Interrogative moves: In an interrogative move, the Inquirer addresses a question to Nature. Nature's answer is added to the left column of the *subtableau* in question. A question naturally must not contain dummy names.

Different kinds of questions:[5] The Inquirer's question can be either a propositional question or a wh-question.

Propositional questions: A propositional question has the form "Is it the case that S_1, S_2, ..., or S_k?" An answer to it is one of the S_i (i = 1, 2, ..., k).

Among propositional questions, there are yes-or-no questions. They are of the form "S_o or not-S_o?".

Wh-questions: A wh-question is of the form "Which individual, call it x, is such that S[x]?"

Presuppositions: Before the Inquirer may ask a question, its presupposition must occur in the left column of the *subtableau* in question. The presupposition of a propositional question is

(1) $(S_1 \vee S_2 \vee ... \vee S_k)$.

The presupposition of a wh-question is

(2) (Ex)S[x].

For yes-or-no questions of the form "A or not-A?", where A is atomic, no presupposition is needed. (The presupposition would be in this case $(A \vee \sim A)$, i.e., a quantifier-free tautology.)

Language: The language used in the game is a finite first-order language L (without function symbols).

Model: Nature's being able to answer questions obviously presupposes that the game is played by reference to some one model M of L. It is normally (but not always) assumed that T is true in M.[6] I shall occasionally speak of worlds instead of models.

3. The importance of strategy considerations

The use of game-theoretical concepts in the definition of my interrogative model is not only an expositional convenience. It serves to call attention to what is probably the most important general methodological advantage of the model. In past studies of various kinds of dialogues, philosophers and linguists have typically formulated their concepts and theses in a way that, in terms of my model, apply to individual moves in the interrogative "game", corresponding to particular utterances in a dialogue or discourse. Examples are provided by speech-act theories,[7] whose very name betrays their conceptual focus; and Grice's conversational maxims.[8] There is a sense in which no such theory focusing on particular "moves" can be fully satisfactory, for from game theory we know that no values ("utilities") can in the last analysis be assigned to individual moves in the game, only to (complete) strategies.[9] In other words, there is no theoretically satisfactory way of relating particular moves to the general ends of the dialogue in question, in the case of my model, to the ends of inquiry. Thus theories of the kind just mentioned are bound to remain unsatisfactory in the last analysis. This fate is especially striking when it happens to a theory which, like Grice's, strives to base its theses on the aims of the kind of dialogue it is applicable to, so that the principles of conversational aptness become special cases of the general principles of rationality. I have elsewhere[10] shown how these limitations of Grice's enterprise are reflected by the details of his conversational maxims.

It is thus a major advantage of my interrogative model that is made possible and indeed focuses on questions of strategy and strategy selection. They will serve, I hope, as special cases of more general questions of dialogical (conversational) strategy. Such questions will be among the focal points of this survey. When I shall speak later in this paper of the "logic" of this or that questioning procedure, I mean in the first place the problems of optimal strategy selection.

4. The richness of the interrogative model

The details needed in the definition of my interrogative games are not likely to make much difference to one's initial impression of the simplicity of these questioning games. Basically, all that happens is that the Inquirer puts questions to Nature one by one and uses Nature's answers to them as additional premises for logical inferences. Can anything interesting be found by studying "dialogues" so utterly simple as such question-answer sequences ?

The answer is that my questioning games, far from being a simple or trivial objects of logical and semantical studies, in reality contain a tremendous wealth of interesting and important theoretical issues. Here I shall briefly identify some of the main logical and semantical perspectives opened by the interrogative model.

5. The fallacy of traditional "fallacies"

First, a partly historical point. My interrogative dialogues are obviously not unlike the questioning games played in Plato's Academy.[11] What is perhaps less obvious — or maybe just forgotten — is that Aristotle formulated his theory of so-called fallacies by reference to such dialectical games. Hence the customary discussions of the traditional fallacies which play such a major role in old-fashioned logic texts are to a considerable extent both historically inaccurate and systematically misconstrued.[12] Some of the central mistakes diagnosed by Aristotle are not at all fallacies in the modern sense of mistaken inferences, but rather violations of the rules of certain questioning games. By offering to us a framework for the systematic study of interrogative processes, my model makes it possible for the first time to put several of the traditional fallacies into a real theoretical perspective.

This can be spelled out in the case of individual "fallacies". For instance, the so-called fallacy of "begging the question", or *petitio principii*, is not a fallacy in our sense at all, for it does not deal with inferences in the first place. As Richard Robinson has shown,[13] it meant (in terms of the interrogative model defined above) for the Inquirer to ask ("petition") the big initial or principal question "C or not-C?" instead of trying to answer it by putting admissible "small" questions to Nature. There is no way of turning such a breach of the rules of my interrogative games to a mistaken inference.

The reason why *petitio principii* violates my game rules is that a question must not be asked by the Inquirer before its presupposition has been established. Of course, the presupposition of the main initial (principal) question is not available to the Inquirer in most cases — indeed, in all interesting cases.

Presuppositions of questions also play a role in other so-called fallacies. The most obvious one is the "fallacy" of many questions, exemplified by questions like

(3) When did you stop beating your wife ?

No argument is needed to show that we are not dealing with mistaken inferences here. What is involved in this "fallacy" is a violation of the requirement that the presupposition of a wh-question must be established before the question is asked. (Cf. section 2 above.)

Similar remarks can be made concerning other so-called fallacies. For instance, the *ignoratio elenchi* can be taken to be a mistake concerning the conditions on which my questioning procedure can be used to refute the initial theoretical premise T.

Thus the interrogative model puts a large part of traditional logico-semantical discussion to a new light by providing the first satisfactory framework for discussing Aristotelian "fallacies".

6. *Questions and answers restricted*

This is nevertheless only a small part of the story. The most important way in which the interrogative model can be seen to yield a veritable embarrassment of logical riches is to note that the description of the model given above is in certain respects incomplete. One of the things that have not yet been specified is under what conditions Nature will actually answer the Inquirer's questions.[14] It is obviously not necessary — nor desirable — to assume that Nature can answer everything that the Inquirer asks her. There are many kinds of possible restrictions, but two dimensions are especially important.

7. *Nature not omniscient ?*

First, Nature might not answer questions for reasons structurally similar to ignorance. Heuristically, Nature may be considered in this case as not being omniscient. Nature will then give only such answers (conclusive answers) as are implied by what Nature knows (or, perhaps, as are in some suitable sense immediately implied by what Nature "knows"). When the interrogative model is used to discuss the scientific process and experiments and observations are conceptualized as questions to Nature such phenomena as limited observational accuracy fall under this heading. This dimension of possible restrictions on Nature's answers, and their implications for the Inquirer's questioning strategies, deserves closer scrutiny, which nevertheless will not be attempted here.[15]

8. *EA-hierarchy of answers*

Instead, I shall briefly discuss another dimension of possible restrictions. They are in terms of the logical complexity of the available answers. (The same restrictions can of course be formulated instead as pertaining to the logical form of the questions to which they are answers.) It can be assumed for simplicity that the answers are in the prenex form, i.e., have the form of a string of quantifiers followed by a quantifier-free expression. The number of changes of quantifier-kind (existential to universal or *vice versa*) will then be one important kind of measure of logical complexity.

More explicitly, I can define a hierarchy of prenexes as follows:

A quantifier-free expression has an $A^o = E^o$ prenex.

A string of existential quantifiers followed by an A^n-prenex is an E^{n+1}-prenex.

A string of universal quantifiers followed by an E^n-prenex is an A^{n+1}-prenex.

This hierarchy of prenexes defines a corresponding hierarchy of formulas and sentences they are prenexes to. I shall call it the EA-hierarchy.

9. *A hierarchy of restraints on answers and its extreme cases*

The types of restrictions which I shall consider in this paper are in terms of the prenexes of admissible answers in the EA-hierarchy just defined. The two extreme cases are the following :

(i) The unrestricted case. In this case, no restrictions are imposed on Nature's answers in terms of the EA-hierarchy. (Of course, this does not exclude other kinds of restrictions.)

(ii) The atomistic case. In this case, only quantifier-free answers are available. It is easily seen that this is to all practical purposes tantamount to allowing only yes-or-no questions concerning the truth or falsity of atomic sentences.

In order to gain some heuristic grasp of these two models, it may be useful to see what kinds of inquiries they can be applied to. The unrestricted case (i) can be thought of as being approximated by a clinical inquiry in a field where a great deal of advanced theoretical knowledge is available to the Inquirer, who is essentially applying it to a new case. Of course, the Inquirer does not have all the relevant knowledge in mind to begin with; it has to be brought to bear on the inquiry by means of suitable questions. Alternatively, the inquiry of a clever detective can be thought of as being of this kind. (Accordingly, I am tempted to call this the "Sherlock Holmes case".) A Sherlock Holmes or Hercule Poirot can appeal both to general truths about every day life ("the only person a trained watchdog doesn't bark at in the middle of the night is its master") and in some rare cases also to general scientific laws, over and above observations concerning particular cases.

10. The atomistic assumption

It is clear, however, that such ready availability of information about general laws is not characteristic of empirical inquiry in science. Mother Nature will not give direct answers to questions which deal with what happens always and everywhere or questions cocerning multiple interactions of all individuals in the world. Indeed, all questions an empirical scientist apparently can expect an answer deal with what happens in particular cases. What we can observe is what is happening here and now in *these* particular circumstances. Hence the atomistic assumption (ii) seems to be the best one to characterize empirical inquiry in science. There are some reasons to think that many recent as well as older philosophers of science have in effect been working on the basis of the atomistic assumption.[16]

11. A spectrum of logics of inquiry

One of the most remarkable features of the interrogative model is that the nature of the questioning process, including the choice of optimal questioning strategies, depends radically on the restrictions which may or may not be placed on available questions in terms of the EA-hierarchy. It might in fact be apter to speak of entirely different models of question-answer dialogues, except that the general framework into which I have placed these various models of inquiry also enables us to compare them with each other. The richness of logical and semantical problems which is generated by my approach is nowhere more dramatically in evidence than in the variety of logics of inquiry arising from different restriction in the EA-dimension.

This observation carries remarkable epistemological suggestions. What it shows is that there is no unique logic of inquiry. The nature of inquiry, including the search for optimal strategies of inquiry, depends essentially on the precise conditions in which the inquiry takes place. Whether this applies also to scientific inquiry remains to be discussed. (Cf. below, end of section 18.)

12. The logic of unrestricted interrogation

First, what is the logic of the "Sherlock Holmes" case, i.e., the logic of inquiry unrestricted in the EA-dimension? For simplicity, let's assume that there are no other restrictions present, either. Then at any stage of the *tableau* construction, there are as many questions the Inquirer can ask as there are (i) disjunctions and (ii) existentially quantified sentences (both without dummy names) in the left column of the *tableau*. For these are then all and only sentences which can serve as presuppositions of answerable questions.

In both cases, the same sentence can be used also as a basis of a *tableau* rule. However, asking the corresponding question is never disadvantageous and frequently advantageous to the Inquirer. For instance, consider the following situation :

(4)	T	C
	—	—
	—	—
	—	—
	—	—
	(Ex) S[x]	
	—	
	—	
	—	

Here the Inquirer can either (i) add to the left column a formula of the form S[α], where "α" is a new dummy name. Alternatively, the

Inquirer can use (Ex)S[x] for the purpose of asking the wh-question whose presupposition it is. The answer will be of the form S[b], where "b" is the name of some member of the domain of individuals do(M) of the model M.

Which outcome is better for the Inquirer ? Obviously the latter, for there normally is more that can be found out about the real individual b through further questioning than about the "arbitrarily chosen" dummy individual represented by "α".

Analogously, it is normally advantageous for the Inquirer to use a disjunction (S_1 v S_2) in the left column as a presupposition of a question rather than as a step in a purely deductive *tableau* construction.

These observations yield a preliminary and approximate but nevertheless important answer to the question: What is the logic of unrestricted ("Sherlock Holmes type") inquiry ? What my observations show that in this case questions (interrogative moves) can always (subject, of course, to other kinds of possible restrictions) replace deductive moves. This makes the overall questioning procedure, which is calculated to model empirical inquiry, structurally similar to the corresponding purely deductive procedure, assuming that the Inquirer is pursuing an optimal strategy. In this sense it can be said that *the logic of unrestricted question-answer dialogues is very close to ordinary deductive logic (proof theory).*

This point can be strengthened further. It is known from proof-theoretical studies that the crucial consideration in deductive strategy selection is the selection of an existentially quantified formula to be instantiated (from the ones that at the given stage of deduction are available for the purpose). But we just saw that these existential instantiations are precisely analogous to the selection of the wh-question to be asked at that stage of the inquiry. Hence the problem of optimal strategy choice is closely similar in the two cases.

The importance of this observation can perhaps be highlighted by recalling that the optimal deductive strategies have been shown not to be mechanical (recursive) in general. Even though I cannot at this time present an explicit formal proof, this result suggests, in virtue of the analogy between deduction and questioning noted earlier, that empirical discovery is not subject to mechanical rules, either, at least not in the unrestricted case.

Another corollary of the partial analogy is a vindication of the traditional idea that deduction, deductive logic and logical inference are valuable tools in substantial empirical inquiry. Even though I have not here cast any doubts on the idea that purely logical inferences are tautological, we have found that the problems of strategy selection are closely related in the purely deductive case and in the case of unrestricted interrogation. Even if this analogy should not be extendible to other kinds of questioning procedures, it serves forcefully to vindicate what I have called the Sherlock Holmes conception of logic and deduction.

13. The role of the subformula property

Even some of the disanalogies that there are between the two procedures, deduction and unrestricted interrogative inquiry, only serve to reinforce the strategic similarity between them. One feature which distinguishes unrestricted questioning procedures from deduction is that the logic of these questioning procedures does not satisfy the subformula property. In other words, no cut-elimination theorem is possible in the unrestricted case. In still other words, being allowed to adjoin tautological premises of the form $S_i \vee \sim S_i$ to the left column of the Inquirer's *tableau* can increase the range of what can be established by means of the interrogative process, depending of course on the selection of the admissible tautologies of this form. As an extreme case, if the tautologies in question are not restricted at all, we obtain a kind of completeness result for the unlimited case.

Completeness Theorem: By means of an unlimited questioning process, with arbitrary tautologies $S_i \vee \sim S_i$ adjoinable to the left column of the game *tableau*, conducted on a model M, any sentence S_o true in M can be established.

Completeness Theorem can easily be proved by induction on the number of logical symbols in S_o.

The reason why the failure of the subformula property does not detach the problem of strategy selection in unrestricted questioning from its counterpart on the deductive side is that the choice of suitable tautological premises $S_i \vee \sim S_i$ plays a crucial role in the search of optimal deductive strategies, too. Even though the same conclusions can on the deductive side be proved without such tautological premises, it is known that their use can shorten deductive proof essentially. Hence the choice of suitable tautologies to be used as additional premises plays an essential role in the search for optimal strategies both in pure deduction and in unlimited questioning, albeit in a different way.

14. *The logic of the atomistic case: model-oriented logic*

The atomistic case produces an entirely different kind of logic. Since I have studied some of its main features elsewhere,[17] suffice it to summarize the main facts.

The logic of atomistic questioning is basically the logic of what is logically implied not by the initial theoretical premise T alone, but by T together with the diagram (state-description) $\Delta(M)$ of the model M by reference to which the questioning procedure takes place. (The diagram $\Delta(M)$ is the set of all negated or unnegated atomic sentences true in M.) Although results concerning questions as to what can be done on the basis of $\{T\} \cup \Delta(M)$ are found in the literature, they have not been brought together and systematized. I have proposed to call a systematic study of the resulting logic *model-oriented logic*. Among the results that belong to model-oriented logic and illustrate its nature, there are the following :

(i) An optimal theory T in the atomistic case is not a complete theory, but a model-complete theory.

(ii) The counterpart to definability in model-oriented logic is identifiability (in the sense used in econometrics). Such identifiability is manifested in the form of definitions whose *definiens* depends on a finite number of members of do(M) (i.e., contains their names, if we are dealing with a particular model).

(ii) One appropriate tool for studying model-oriented logic are constituents and normal forms.[18] In model-oriented logic we study what happens at successively deeper-lying layers of constituents, whereas in "normal" logic (e.g., in the study of "normal" definability) we study what happens at the surface of constituents.

Not unexpectedly, the study of the atomistic case can also be brought to bear on a variety of important issues in the philosophy of science.

15. *Atomistic case and explicit definitions*

The properties of model-oriented logic would deserve further study. For instance, in this extreme case, the subformula property (cut elimination) does hold. However, another familiar logical principle goes by the board, viz. the noncreativity of explicit definitions. This result is not surprising in itself. The introduction of a new primitive predicate P by an explicit definition

(5) $(Ax)(Px \leftrightarrow D[x])$

has the effect of turning the complex predicate D[x] into a simple one and hence making questions concerning its applicability to particular cases answerable by Nature, even though they previously could not be answered by Nature in virtue of the atomistic assumption.

In spite of this naturalness of the result, it helps to shed interesting light on the nature of definitions and their role in scientific inquiry.

16. Revisiting the spectrum of interrogative logics

One of the most important features of the overall conceptual situation is nevertheless that the atomistic case with its "model-oriented logic" is only one extreme in a long spectrum of essentially different logics of questioning (inquiry). It is hence possible and indeed interesting to raise questions concerning the intermediate logics of questioning, for instance, the question as to in which of them the subformula property or the noncreativity of explicit definitions holds good. On the level of general philosophical issues, the existence of the spectrum suggests, even though it does not yet prove it, that philosophers' quest of *the* logic of inquiry or *the* logic of scientific discovery is a pursuit of a chimera. What the appropriate logic of an inquiry is depends (among other things) on its position in the EA-spectrum of restrictions on available answers to the Inquirer's questions.

17. The inductivist case

What substantiates this conjecture and makes the study of the spectrum of logics of questioning even more interesting is that some of the intermediate cases have a great deal of intrinsic interest. For instance, the classical inductivist conception of inquiry can be taken to amount to saying that Nature can also give to the Inquirer A^1-answers, over and above A^o-answers. For the gist of the inductivist position is that by examining a sufficient number of particular cases an Inquirer can establish whether an empirical generalization from such particular cases holds or not. The inductivist conception can perhaps then be criticized by trying to show that this A^1-restriction does not take us essentially further than the atomistic restriction.

18. A^2 logic as the logic of experimental inquiry

Most importantly, it can be argued that some of the intermediate cases are much more realistic models of experimental inquiry than either the unrestricted or the atomistic extreme case. For what is it that a scientist can find out by means of a controlled experiment? Not just information about particular cases. What a scientist is typically trying to do in such an experiment is to vary one variable characteristic of the experimental situation and by observing another variable to find out how it depends on the controlled variable. In other words, the answer that a scientist can receive from a controlled experiment, thought of as a question put to Nature, is a functional dependence between two or more variables. Now the expression of such a dependence is not an atomic sentence nor an A^1 or E^1 sentence. It has at least the logical complexity of an A^2 sentence. Hence the true logic of experimental science is not model oriented logic or deductive logic. It is A^2 logic, which therefore comes as close as anything in my spectrum to the true logic of scientific inquiry in advanced experimental sciences.

Now what is A^2 logic like ? The only honest answer is that apparently nobody knows. There are relevant results in the logical literature, but there does not seem much of a comprehension of the overall situation comparable, e.g., with what is known (and understood) of the atomistic case.

It can nevertheless be seen that the logic of the A^2 case differs sharply from most philosophers' ideas of the scientific method, which therefore stand in need of critical scrutiny.

The main point is clear. Whereas no complex nested-quantifier conclusions can be derived by means of my interrogative processes (including deduction) from atomistic answers or A^1 answers, without essential reliance on an initial complex general theory T, such conclusions are possible at least in some situations for A^2 answers. This puts to a new light many things in the history and philosophy of science, among them Newton's claim of having "derived" or even "deduced" general laws from "phenomena". This claim, which most philosophers of science have been unable to make much sense of, now receives a perfectly natural interpretation. What Newton is saying is that general laws can be derived from answers to A^2 questions, i.e., derived by analysing phenomena by means of controlled experiments. This account fits in well both with Newton's emphasis on the use of analysis, i.e., the study of functional dependencies between different factors, and with his keen awareness of the nature and role of controlled experiments.

This is only one example of the power of A^2 logic to throw new light on different kinds of inquiry (different kinds of questioning processes). More study of the logical situation is nevertheless needed as a basis of such applications.

It is also in order to note that for non-experimental sciences, for instance, for purely observational sciences, different EA-restrictions and hence a different logic of inquiry may be appropriate, and likewise for purely clinical investigations within an applied science (e.g., medicine).

19. Inquiry and dialogue

Thus the apparent simplicity of my questioning model turns out to cover a tremendous variety of different structure and applications. Some of the applications sketched in this paper are more relevant to a scientist's dialectical interaction with nature than to dialogues between human speakers. The reason is that restrictions of replies to atomistic or A^1 or A^2 answers is more characteristic of a scientist's situation vis-à-vis nature than of what one is likely to face in a conversational situation in everyday life. But this is merely a matter of degree, as the example of Socratic dialogues vividly shows. (There is no reason why Socratic or Aristotelian questioning dialogues could not be restrained, e.g., by the atomistic assumption.) Indeed, Socratic *elenchus* is likely to be one of the test cases that first occur to a philosopher. Hence the rarity of such dialogues in the rough-and-tumble modern ordinary discourse does not reflect on the significance of the question-answer procedures codified in my interrogative games also as a case study designed to illustrate the richness with a general theory of dialogues will have to cover.

Florida State University

NOTES

*The work that resulted in this paper was made possible by NSF Grant no. IST-8310936 (Information Science and Technology, Principal Investigators Jaakko Hintikka and C.J.B. Macmillan).

[1] Along different lines, the same strategy has been followed by Lauri Carlson in his important book *Dialogue Games: An Approach to Discourse Analysis*, D. Reidel, Dordrecht, 1983.

[2] These question-answer dialogues and/or certain similar ones have been studied in my earlier papers, including the following : "The Logic of Science As a Model-Oriented Logic", in P. D. Asquith and P. Kitcher, editors, *PSA 1984*, Philosophy of Science Association, East Lansing, MI, 1984, pp. 177—185; "Rules, Utilities, and Strategies in Dialogical Games", in Lucia Vaina and Jaakko Hintikka, eds., *Cognitive Constraints on Communication*, D. Reidel, Dordrecht, 1984, pp. 277—294; (with Merrill B. Hintikka) "Sherlock Holmes Confronts Modern Logic", in E. M. Barth and J. L. Martens, eds., *Theory of Argumentation*, Benjamins, Amsterdam, 1983, pp. 55—76; "The Logic of Information-Seeking Dialogues: A Model" in W. Becker and W. Essler, eds., *Konzepte der Dialektik*, Vittorio Klostermann, Frankfurt am Main, 1981, pp. 212—231.

[3] For the *tableau* method, see E. W. Beth, "Semantic Entailment and Formal Derivability", Mededelingen van de Koninklijke Nederlandse Akademie van Wetenschappen, Afd. Letterkunde, N.R., vol. 18, no. 13, Amsterdam, 1955, pp. 309—342; reprinted in Jaakko Hintikka, ed., *Philosophy of Mathematics*, Oxford U.P., 1969, pp. 9—41.

[4] The usual basic rules always introduce subformulas of earlier ones. The others can all be reduced to the introduction of tautological premises of the form (S v ~S) into the left column, where the relation of S to earlier entries in the *tableau* is not restricted in any way.

[5] For a systematic discussion of different kinds of questions, of the question-answer relation, and of the concept of presupposition, see Jaakko Hintikka, *The Semantics of Questions and the Questions of Semantics*, Acta Philosophica Fennica, vol. 28, no. 4, Helsinki, 1976.

[6] When this assumption is relaxed, we can study, for instance, how T can perhaps be refuted by means of the questioning process.

[7] See, e.g., John Searle, *Speech Acts*, Cambridge U.P., 1969; Peter Cole and Jerry L. Morgan, eds., *Speech Acts (Syntax and Semantics*, vol. 3), Academic Press, New York, 1975.

[8] Paul Grice, "Logic and Conversation", in Cole and Morgan, note 7 above, pp. 41—58.

[9] The outcome of a game (the payoffs of different players) is not determined in general before the strategies of all players are determined. Conversely, the outcome is uniquely determined when each player has chosen a pure strategy, and the expected outcome is determined when each of the players has chosen a mixed strategy.

[10] Jaakko Hintikka, "Logic of Conversation as a Logic of Dialogue", in R. Grandy, editor, *Philosophical Grounds of Rationality: Intentions, Categories and Ends — A Festschrift for Paul Grice*, Oxford U.P., 1985, pp. 257–274.

[11] For a spirited description of these games, see Gilbert Ryle, "Dialectic in the Academy", in R. Bambrough, ed., *New Essays on Plato and Aristotle*, Routledge and Kegan Paul, London, 1965.

[12] For an excellent survey of the traditional fallacies, see C. L. Hamblin, *Fallacies*, Methuen, London, 1970.

[13] Richard Robinson, "Begging the Question 1971", *Analysis* vol. 31 (1971), pp. 113–117.

[14] Another thing that remains to be specified more closely is the determination of payoffs. For some remarks on the problem, see my papers, "Rules, Utilities, and Strategies in Dialogical Games" (note 2 above).

[15] The main upshot is to turn my questioning procedures into a quest of "interpolation formulas" in logicians' sense. This does not disassociate the procedures from deductive logic in the least.

[16] For instance, both inductivist and hypothetico-deductive theories of science usually assume that a theory can be confronted with experience only in terms of particular consequences of the theory.

[17] See "The Logic of Science as a Model-Oriented Logic", note 2 above.

[18] For them, see, e.g., Veikko Rantala, *Aspects of Definability*, Acta Philosophica Fennica, vol. 29, nos. 2–3, Helsinki, 1977.

7

WHAT IS THE LOGIC OF EXPERIMENTAL INQUIRY?*

1. THE INTERROGATIVE MODEL

The purpose of this paper is to expound one of the most important insights yielded by the interrogative model of inquiry, prominently including scientific inquiry. I have outlined this model elsewhere.[1] The basic idea on which this model is based is simplicity itself. It can be expressed most easily in the jargon of game theory.[2] A player, called *the Inquirer*, is trying to prove a predetermined conclusion C from a given theoretical premise T. (In a variant form, the Inquirer is trying to prove either C or $\sim C$, i.e., to answer the initial question "C or not-C?".) Over and above *deductive moves*, i.e., over and above drawing logical inferences, beginning with T, the Inquirer may address questions to a source of information and use the answers (when available) as additional premises, in short, may carry out *interrogative* moves. The answerer is called *Nature*. As a bookkeeping technique, a Beth-like semantical tableau can be assumed to be used.[3] Each move is relative to the stage of a subtableau reached in the game at the time. The questions must of course pertain to a given model M of the language of T. Before a question is asked, its *presupposition* must have been established by the Inquirer, that is, occur in the left column of the subtableau in question.[4] The deductive moves are restricted to those stages of tableau construction that satisfy the subformula principle, except for a given set of tautologies ($S_i \vee \sim S_i$) that may be inserted as extra premises. (It is also assumed that the tableau rules do not move sentences from one column to the other.)

Those explanations do not yet specify completely any one interrogative game, only a family of different kinds of games. It quickly turns out that the most important further specifications are the restrictions (or lack thereof) that can be imposed on Nature's answers. Of these sundry restrictions, those that govern the maximal admissible logical complexity – more specifically, the complexity of the quantifier prefix – of Nature's answers are among the most interesting ones. There is a hierarchy of such restrictions defined by the maximum number of admissible quantifier-kind changes in the prefix of an answer. I shall call it the *AE*-hierarchy. Thus an A^n-prefix is a string of universal quantifiers followed by an E^{n-1}-prefix, and an E^n-prefix is a string of existential quantifiers followed by an A^{n-1}-prefix. An $A^0 = E^0$ prefix as an empty one, i.e., the "prefix" of a quantifier-free expression. A restriction to the $A^0 = E^0$ case turns out to be equivalent to restricting Nature's answers to negated or unnegated atomic propositions. Accordingly, the assumption that Nature's answers are restricted to $A^0 = E^0$ ones will be called the *Atomistic Postulate*.

2. TYPES OF INQUIRY

Why is this an important hierarchy of interrogative games? A part of the answer is that we can characterize different actually occurring types of inquiry and distinguish them from each other by locating them in this *AE*-hierarchy. The niche a type of inquiry occupies in this hierarchy determines its logic in an eminently natural sense of the word. For instance, clinical inquiry, say a physician's search for a medical diagnosis or a detective's "deduction", is (partially) characterized by the absence of structural restrictions.[5] As a part of her or his reasoning, a physician may look up general medical information in a handbook, and a Sherlock Holmes may bring his extensive knowledge of the general truths of chemistry to bear on his "deductions", no matter what the logical complexity is of the truths evoked.

What characterizes clinical inquiry, therefore, is not that it does not aim at general laws, but that it utilizes previously established but initially tacit generalizations. Indeed, sometimes a general result is in fact established via clinical inquiry. In terms of the interrogative model, this depends merely on what the intended consequence C is, not on the logical structure of the inquiry. Frequently a clinician's diagnosis in fact has some degree of generality; witness, e.g., a physician trying to establish the cause of an epidemic.

For another example, in recent linguistics it has been widely assumed that the basic data are a competent speaker's intuitions concerning particular sentences (strings).[6] In the framework of the interrogative model, this means that all the answers that the Inquirer *qua* linguist can hope to receive are quantifier-free positions, i.e., $A^0 = E^0$ propositions. Yet no really satisfactory rationale for this assumption has ever been given. Is a competent speaker of English any less competent to specify, say, how the plural of a noun phrase is formed than to specify whether some long and complicated string is grammatical or not? Indeed, the Chomskian assumption of innate general ideas concerning grammar makes it unlikely that a competent speaker's intuitions are limited to particular cases. Yet giving up the Atomistic Postulate would make a drastic difference to the entire logical structure of contemporary linguistics. The example of linguists' methodology therefore shows how important it may be to locate a type of inquiry in the *AE*-hierarchy.

3. WHAT IS THE LOGIC OF EXPERIMENTAL SCIENCE?

The most important issue here is nevertheless the question of what the logic of a typical empirical natural science is, i.e., where the kind of inquiry exemplified by such empirical scientific inquiry is located in the *AE*-hierarchy. One's first impulse here is undoubtedly to say that such inquiry is characterized by the Atomistic Postulate. Surely good old mother nature will never tell a scientist directly what the case is always and everywhere. The only answers she is going to yield to an inquirer concern particular cases, for instance, what one particular experiment or observation can show to the Inquirer. This restriction is of course precisely what I have called the Atomistic Postulate.

It is fairly obvious that the logic of purely observational sciences is in fact defined by the Atomistic Postulate. However, this does not apply to all natural sciences. The two main theses that can be argued for here are the following:

(1) The logic of experimental (as distinguished from a purely observational) science has at least an A^2-complexity.

(2) In virtually all of the recent philosophy of science, the Atomistic Postulate has been assumed.

The potential interest of these theses can be seen from the fact that together they imply that virtually all recent philosophy of science needs a major overhaul. The two theses, especially the latter one, deserve a fuller discussion than they can be given here. In fact, I will restrict my comments in this paper to thesis (1). At one time – about twenty or twenty-five years ago – it was fashionable to speak of the dogmas of the philosophy of science. In that jargon, the Atomistic

Postulate is the super-dogma of recent philosophy of science, if I am right.

4. AN ANALYSIS OF A CONTROLLED EXPERIMENT

First, consider a controlled experiment. This is in fact the truly classical application of the interrogative model. Controlled experiments were considered as paradigmatic examples of "questions put to nature" in Kant's famous statement of the idea on which the interrogative model is based.[7] What happens in such an experiment? The very idea of a controlled experiment is that the scientist can vary, within certain limits, one variable (the controlled variable) and observe how another variable (the observed variable) changes accordingly. The outcome of a controlled experiment is hence a statement of the dependence between the two variables. What will such a statement look like? It will say that for each value of the controlled variable there corresponds such-and-such a value of the observed variable. An answer to a question put to nature is therefore of the A^2-form.

A question with this kind of answer could be expressed in a natural language by a locution like "Whom (what) does everyone . . ."?[8] The presupposition of such a question is of the form

(1) $(\forall x)(\exists y)S[x, y].$

An answer will specify a function, say f. As I have shown elsewhere (op cit., note 8), the desideratum of the question will be of the form

(2) $$\left.\begin{array}{c} K \\ \nearrow \quad \searrow \\ (\exists y) \leftarrow (\forall x) \end{array}\right\} \quad S[x, y]$$

where "K" is the formal counterpart to "the Inquirer knows that". Here (2) is equivalent with

(3) $(\exists f)K(\forall x)S[x, f(x)].$

This form (3) of the desideratum brings out especially clearly the fact that what is being sought for is the function expressing the dependence of one variable or another.

Then f will be a conclusive answer if and only if

(4) $(\exists h)K(f = h)$

just as in the case of answers to simple wh-questions. Nature's answer will take the Inquirer essentially from (1) to

$$(5) \qquad (\exists f)(\forall x)S[x, f(x)].$$

This step is mathematically speaking an application of the principle of choice, except that Nature's answer provides the Inquirer with a specific known function f. Earlier, I have shown that answers to simple wh-questions are analogous to existential instantiations and answers to propositional questions are analogous to applications of tableau rules for disjunction.[9] The same analogy is now seen to hold between A^2-answers and the principle of choice. (What this tells a philosopher of logic about the principle of choice, I leave to the reader to decide.)

5. THE STRUCTURE OF MORE COMPLICATED EXPERIMENTS

Proposition (1) can be thought of as a description of the experimental situation. There "x" is the controlled variable and "y" the observed variable. By considering descriptions of more complicated experimental situations, it can be seen that a controlled experiment can yield answers more complex than A^2-propositions.

For instance, assume that in an experimental setup we have two controlled variables x and z and two observed variables y and u, of which y depends on x and u depends on both x and z. Then the proposition

$$(6) \qquad (\forall x)(\exists y)(\forall z)(\exists u)S[x, y, z, u]$$

can serve as the presupposition of a question whose desideratum can be represented as

$$(7) \qquad (\exists f)(\exists g)K(\forall x)(\forall z)S[x, f(x), z, g(x, z)]$$

or as

$$(8) \qquad K(\forall x)(\exists y)_K(\forall z)(\exists u)_K S[x, y, z, u]$$

where the subscript "K" indicates a quantifier's independence of this epistemic operator. Answers will then be of the form

$$(9) \qquad (\forall x)(\forall z)S[x, f(x), z, g(x, z)]$$

where f and g have to satisfy certain conditions. This is more powerful logically than (6), being obtainable from (6) by repeated applications of the principle of choice. Hence some answers to those questions to nature which involve controlled experiments are stronger than A^2-propositions.

This line of argument depends on the assumption that the experimenter knows which variables depend on which. The status of such assumptions deserves further investigation. It seems to me that such dependence and independence assumptions are legitimately made all the time in science. Even if they are, logically speaking, contingent, and sometimes problematic (witness quantum physics with its non-locality phenomena), in some cases no serious scientist will doubt them. For instance, no one expects in this day and age that the electromagnetic system interacts with the gravitational one, at least not on the level of terrestrial macrophenomena.

But if such independence assumptions are legitimate, the results of experiments can logically speaking be extremely strong. Consider, for example, an experimental situation in which an observed variable y depends on controlled variable x alone, and u likewise on z alone, but in which all the four variables interact. Then the presupposition of the relevant question is the branching-quantifier sentence[10]

$$(10) \qquad \begin{array}{l} (\forall x)(\exists y) \\ (\forall z)(\exists u) \end{array} \!\!> S[x, y, z, u].$$

The desideratum is of the form

$$(11) \qquad K < \begin{array}{l} (\forall x)(\exists y)_K \\ (\forall z)(\exists u)_K \end{array} \!\!> S[x, y, z, u]$$

which is equivalent with

$$(12) \qquad (\exists f)(\exists g)K(\forall x)(\forall z)S[x, f(x), z, g(z)].$$

The answers will be of the form

$$(13) \qquad (\forall x)(\forall z)S[x, f(x), z, g(z)]$$

together with additional conditions with which (13) prefixed by "K" implies (12), and hence *a fortiori* (10).

Now propositions of the form (13) are extremely strong, so strong that this logic comes close to that of the entire second-order logic.[11] Hence extremely strong propositions can apparently be forthcoming as results of controlled experiments.

This logical strength is due much more to independence assumptions, e.g., such assumptions as enable us to represent the experimental situation as (10) rather than as (1). In fact, the presupposition (10) is already extremely strong, and the step from it to the desideratum (12) or to (13) does not seem to add so very much extra force, relatively speaking.

Be this as it may, what has been found shows that the logic of experimental inquiry can be stronger than that of A^2-inquiry, even though the extent and nature of the extra force requires further study.

6. THE ROLE OF EXPERIMENTS IN INQUIRY

One may perhaps object that it is not correct to look at what happens in a controlled experiment as an A^2-question and answer, since the primary data yielded by an experiment and typically particular recordings of the registrations of measurement instruments. In the simplest cases, these are pairs of values of the controlled variable (argument values) and of the observed variable (function values). We can think of them as marked in a coordinate system. The further step of plotting a curve through these points as closely as possible and of finding a mathematical expression whose graph matches this curve may be alleged to be a further inferential step. According to this line of thought, the primary data must hence be expressed by particular propositions, just as the Atomistic Postulate requires.

To this objection one can reply in several complementary ways. First, it is beneficial to realize what we are trying to ascertain here. What is at issue is not the sociology of science but its logic. I am not dealing with what scientists say of their work or what they write in their research reports. What counts here is the role which the results of controlled experiments play in the entire scientific process. This means in practice asking what significance results of controlled experiments have in further scientific reasoning. And it is my thesis here that the results of controlled experiments enter into this further reasoning in the form of functional dependencies. This can in fact be considered the characteristic mark of experimental, as distinguished from purely observational, reasoning.

In this respect, the concept of experiment is like the concept of observation which I have discussed separately.[12] What characterizes this concept and what is intended by such phrases as "the concept-ladenness of observations" and "the theory-ladenness of observations" is the role observations play in scientific inquiry, not, e.g., some psychological interaction between one's concepts and one's observations.[13] By a similar token, I am arguing here that the results of controlled experiments enter into a scientist's "game of exploring the world" (Ryle's phrase) via the functional dependencies they reveal.

7. AN EXAMPLE

It is in any case clear that in many important actual historical cases the results of experimental inquiry enter into the later stages of scientific reasoning precisely as I have indicated, viz., in the form of functional dependencies, typically expressible as mathematical laws, which are then generalized or reconciled with other similar laws or otherwise used as stepping-stones for further theoretical results.

A well-known case in point is Planck's discovery of the idea of the quantum, i.e., a minimal separable amount of energy.[14] His problem situation consisted in the study of so-called "black body radiation", that is, radiation emitted by a completely absorbent body. More specifically, one of the questions Planck faced was the relation of the temperature T, frequency of radiation ν, and the energy density per unit interval of wavelength u_ν. The problem was not that Planck or his comtemporaries did not have experimental results concerning their interrelation. Rather, it was that he had too many experimental results. For extremely low frequencies there was the so-called "Rayleigh-Jeans radiation law"

$$(14) \qquad u_\nu = \frac{8\pi\nu^2 kI}{c^4}$$

However, for high frequencies this law was known to fail. In its stead was another law, Wien's radiation law:

$$(15) \qquad u_\nu = \alpha\nu^3 \exp(-\beta\nu/T)$$

which in turn failed for low frequencies. The fact that these laws had been arrived at by means of theoretical arguments before they were experimentally confirmed does not matter for my purposes. The two laws (14)–(15) could in principle have been arrived at purely experimentally, and as a matter of historical fact nobody seemed to consider the theoretical arguments leading to (14)–(15) as anything more than ad hoc heuristic ones.

Planck's problem was to reconcile the two laws (14)–(15). This is no mean feat, in view of the fact that (14) and (15) are quite different functions mathematically. How can they be interpreted as special (extreme) cases of one and the same deeper mathematical law? What Planck found was that the only reasonable way of doing so was to assume the existence of the quantum. His line of reasoning and its stringency (or lack thereof) does not concern us here. What is relevant for my current purposes is the nature of the reconciliation. This reconciliation meant a reconciliation of the two different mathematical laws, not of the different particular observations (triples of values of variables) that went into the experimental establishment of these laws. In order to reconcile these particular observations, all that would have been needed is to plot a curve (strictly speaking, a surface, because we are dealing with two-place arguments) through the points representing particular measurement results. Anybody can do that; no new insights are forthcoming from such mechanical interpolation and extrapolation. In contrast, the problem Planck faced was to reconcile different mathematically expressed laws of a limited scope. And this means that results of experiment entered into Planck's reasoning in the form of laws and not as sets of particular measurement results.

8. EXPERIMENTS VS. OBSERVATIONS

I am not saying, of course, that all the data of an experimental science – i.e., all answers to questions which the Inquirer addresses to Nature – are of the A^2 or functional dependence form. In a rough-and-ready way, we can distinguish from each *observational* move, whose result is an atomistic answer, *experimental* moves, whose result is an A^2-proposition. This terminology has to be taken with a large grain of salt, however. There are important controlled experiments, for instance some experiments carried out by means of particle accelerators, whose results are particular measurement results, "points on the graph paper", so to speak. I do not want to argue, either, even though it would be viable to do so, that a philosopher should consider such experimental setups, however complex and highly controlled they may be, as merely devices for creating a setting for suitable observations. The point of my terminology can be appreciated without such excuses.

Conversely, if what we normally call observations can establish interesting functional dependencies, they can serve the same purpose as controlled experiments in my idealized model. An example is provided by Kepler's laws, which of course were established by what everybody would call observations, but which do establish laws tying together time and various parameters characterizing the movements of a planet.

A little clarification of terminology and history is perhaps helpful here. Often, philosophers lend the term "experiment" a much wider sense than I have done here and call any creation of an observational situation which would not have come about without a scientist's intervention an "experiment". I don't propose to change this terminology, but I do want to call attention to a narrower sense of the term which might be labelled "controlled experiment" or (to borrow an apt term from O. Becker)[15] "analytical experiment" and which is the sense in which I am employing the word here. It is important to pay attention to this narrower concept, for it makes a tremendous difference, for instance, to our understanding of the methodology of early modern science. In what sense is it that we can speak of it as relying on experiments? T. S. Kuhn has contrasted what he calls mathematical and experimental traditions in the development of physical science, using the word "experimental" in an undifferentiated sense.[16] From the vantage point of the narrower concept of experiment, this is a false contrast, for the very point of a controlled experiment is to find out a functional dependence, ideally expressed as a mathematical law. The entire contrast between the experimental and mathematical traditions thus overlooks a crucially important bridge between the ideas of experiment and of mathematization of science. This link may even be what is truly novel in the methodology of early modern science. In fact, several scholars have seen in the idea of an analytic (controlled) experiment an important focal point of the methodology of early modern scientists. In the light of such an idea, we can, for instance, understand the crucial role played by the analytic method (which can be traced back to ancient Greek geometers) in early modern science and mathematics.[17] From this perspective, Newton emerges as the clearest representative of all early modern science methodologically, and not as the anomaly which Kuhn has to brand him as.

9. SCIENTIFIC INQUIRY IS MANY-LEVELLED

Another partial answer to possible criticisms is based on the following idea: scientific inquiry does not move on one level only. When I am talking of the way in which the results of a controlled experiment enter into the scientific process, I am thinking of one level of inquiry, a level on which an experiment operates as a single direct question to nature. It is not incompatible with this idea to consider the experiment itself as a complicated process which moves on a different level, which I will call, tentatively, the micro-level. Everybody who has been involved in a complex physical experiment or comes to appreciate one knows that they typically involve an elaborate technology, in our day and age normally electronic technology, which plays the role of a "theoretical" premise. I will not try to answer here the question how such micro-inquiry can be illuminated by the interrogative model.

The interrogative model is in the first place calculated to apply to the macro-level of scientific inquiry. It is nevertheless in order to comment briefly on the structure of micro-inquiry. An essential part of such an inquiry is captured by an analogy which is often used to discuss scientific inference in general. It is in many ways a beautiful and instructive analogy. In it, the task of scientific inference is compared to the problem of drawing a curve that best fits a number of points in a plane. In spite of the instructiveness of the curve-fitting idea, it does not help to understand the macro-structure of scientific inquiry, nor do the vagaries which are associated with curve-fitting affect the structure of macro-level inquiry. For the curve-fitting idea belongs squarely to what I have construed as the micro-level of scientific inquiry.

10. THE LIMITATIONS OF THE CURVE-FITTING IDEA

This point is worth spelling out more fully. We can now answer the objection mentioned above, to the effect that nature's direct answers should be construed as measurement results and not as functional dependencies. This objection is based on a misleading way of assimilating different levels of inquiry to each other. The measurement results may be nature's answer in the micro-level game that an experiment can be thought of as being, not in the macro-level game in which an entire controlled experiment is but a single move.

Essentially the same point can also be made in epistemological terms. A possible objection to what I have said is that particular measurement results should be thought of as nature's direct answers because the passage from them to the general law they instantiate is an inferential step which cannot be immediately certain, and which therefore introduces an uncertainty factor into the rest of a scientist's reasoning. My answer is that this may indeed be the epistemological situation in a micro-level mini-game consisting of one experiment. However, this fact does not force us to think of the results of experiments (that is, according to my account, functional dependencies) as not being certain, as far as the macro-level game of inquiry that the interrogative model is calculated to capture is concerned.[18]

This fact could be illustrated by a variety of examples. It simply is not the case that a scientist considers even a mathematically expressible law emerging from an experiment as something he or she cannot take for granted in further inquiry. This fact was registered by as early a thinker as Leibniz, who compared micro-inference to solving a cryptograph rather than to curve-fitting: "It may even turn out that a certain hypothesis can be accepted as physically certain if, namely, it completely satisfies all the phenomena which occur, as does the key to a cryptograph."

The logico-philosophical basis of such certainty is the subject for a chapter in the study of experimental micro-games. One possibility here is to try to think of inductive logic, more specifically, the logic of inductive generalization, as being an essential part of the logic of those experimental-level games. Inductive logic would thus find a spot, albeit a modest one, on the overall map of scientific inquiry.

Both the historical and the logical aspects of the level distinctions and their consequences deserve further attention.

11. EXPERIMENTAL RESULTS IN FURTHER INQUIRY

I am not saying, of course, that the generalization involved in the passage from measurement results to a functional dependence (i.e., in the simplest cases, the plotting of a curve through the measurement points and the mathematical interpretation of this curve) must be historically and organizationally inseparable from the measurement process. But this does not make any difference to my line of argument here, which concerns the logical structure of scientific inquiry and not its social organization. In fact, even in cases where the generalization is historically a separate phase of the scientific process, the general perspective offered by the interrogative model applies perfectly well. For instance, consider the study of the hydrogen spectrum in the late nineteenth century.[20] After Ångström had measured the wavelengths of the first four hydrogen lines, an obscure Swiss numerologist called Balmer noted that they could be expressed by the formula

(16) $\lambda = hm^2/(m^2 - 2^2)$

where $h = 3645.6 \times 10^{-7}$ and $m = 3, 4, 5, 6$. (λ is assumed to be expressed in millimeters.)

What is the significance of Blamer's observation (which was soon extended experimentally to other hydrogen lines)? What role did it play in subsequent developments? We know the answer: Balmer's formula was the most important experimental result on which Bohr's model of the hydrogen atom was based. In other words, Balmer's formula played precisely the role of an answer to an A^2-question.

What is also relevant here is that the alleged epistemological uncertainty of the outcome of the generalization involved in Balmer's formula never played any role in the further developments which aimed at explaining the hydrogen spectrum. Here we have to distinguish sharply the question of the accuracy of Balmer's formula from the question of the reliability of the generalization it involves. It was known right from the beginning that the formula was subject to slight discrepancies. Explaining these was an interesting task for further research. But even if there had not been any discrepancies, the epistemological problem would have remained, if the Humean inductive problem were a genuine one. And on this alleged problem there is no trace in the actual scientific discussion or research.

This can be generalized. There is a sense in which the traditional problem of induction ("Hume's Problem") is a philosophers' invention which has little connection with the actual practice of experimental science. Of course, there is on the micro-level a process which is very much like the process of inductive generalization, as was indicated. (Balmer's formula can indeed be thought of as being obtained from Ångström's data by some suitable "logic of induction".) But the alleged uncertainty of this process does not matter in the least for scientific practice.

12. UNCERTAIN ANSWERS DON'T CHANGE THE SITUATION

I am not saying, either, that the most realistic version of the interrogative model is the one in which Nature's answers have to be thought of as certain, not merely true with a certain probability. Indeed, the task of developing further the interrogative model so as to allow merely probable answers by Nature is an urgent one. But this further development of the interrogative model will not make a difference in what I have said. This is illustrated by two observations. First, the theory of interrogative inquiry with uncertain answers which is beginning to emerge is quite different from mainstream inductive logic. It still deals with direct inference, and is therefore essentially different from situations considered in traditional discussions of induction. It deals with certainty-preserving inferences from uncertain premises, not with uncertain inferences from indubitable premises.

Second, the really striking fact here is the richness and interest of the interrogative model in which all Nature's answers are assumed to be indubitable. What this means is that even if Hume's problem, the problem of induction, were solved for good, a scientist and a philosopher of science would still face a rich assortment of interesting and important problems, problems incomparably more germane to the actual methodology of sciences than any philosophical query concerning the justification of induction. Against this general background, it is only to be expected that treating the results of controlled experiments as being (typically, not always) functional dependencies does not mean ignoring any appreciably significant problem in the philosophy of science.

13. THE CONCEPT OF EXPERIMENT AND EROTETIC LOGIC

In the last analysis, the most convincing reason why a controlled experiment must be construed as an A^2-question put to Nature lies in the logic of the situation. If the question asked by the experimenter, tacitly or in so many words, is

(17) What value of the observed variable corresponds to each value of the controlled variable?

Then it is is simply a fact of the logic of such questions that each of their answers will have to specify a function. Hence, if it is at all feasible to think of a controlled experiment as a question of the form (17) put to nature, then what has to enter into the rest of the process as the answer to (17) is a function. Thus the functional nature of the outcomes of experiments is an integral part of the very idea of scientific inquiry as a series of "questions put to nature".

However, erotetic logic at the same time shows that there can be natural exceptions to this rule.[21] For it is possible to construe the desideratum of a question like (17) to be of the form

(18) $(\forall x)(\exists y)KS[x, y]$.

Then an answer might consist in specifying the class of pairs of corresponding values of the two variables. However, it can be shown that (18) is just not how we usually construe the desideratum of a question like (17). The reason is that, on the construal (18) of the desideratum of (17), the answers will have to assume, or provide collateral information to guarantee, that the questioner will end up knowing what all the relevant values of the variables are. The unnatural character of such a construal is perhaps seen most clearly from who-questions like

(19) Who is the role model of each younger brother?

In order for the reply "his eldest brother" to be a satisfactory (conclusive) answer to (19) *on the construal* (18) *of its desideratum*, the questioner would literally have to know who everybody and his brother are, or, strictly speaking, who every younger brother and his eldest brother are. This is of course not what is usually aimed at in a question like (19). What is typically required is that the questioner knows (or is brought to know) what the *function* is that is offered as an answer, not what all the terms are that it relates to each other. In the case at hand, this would mean knowing what the function "the eldest brother of *x*" is, which is not a tall order at all. Hence functional answers are typically what is expected as replies to a question like (19).

However, it is in principle possible to understand a natural-language question like (19) as having a desideratum of the form (18), and occasionally it is the intended reading of a query of the form (19). This is the precise logical rationale of the possibility of construing some experiments merely as a series of glorified observations in artificial situations.

Here it seems especially clear that my thesis of the functional nature of Nature's answers to experimental questions is not primarily a sociological or historical thesis. What it does is to spell out, by means of results from the logic of questions and answers, what is entailed in the general idea of construing experimental inquiry as an interrogative process. It amounts to pointing out that the functional character of the outputs of controlled experiments is an eminently natural, albeit not inescapable, feature of this way of looking at experimental inquiry. It is, with the qualifications implicit in what I have said, a conceptual (logical) and not an empirical thesis. It can be accepted or rejected only by accepting or rejecting the entire interrogative model.

Of course the main aim of the interrogative model is to capture the structure of the scientific enterprise. It is for this reason that one of the crucial issues is bound to be the role different "questions put to Nature" and their answers play in the overall process of actual science. This issue is illustrated amply by the earlier parts of this paper.

This concern with the structure of actual scientific reasoning does not diminish the conceptual character of my own inquiry. In fact, it seems to me that we have here an instructive case study of the ways in which a conceptual inquiry – in the present case, mainly erotetic logic – can provide a useful framework for raising questions about the argumentative structure of actual science and placing the answers into an interesting perspective. In other words, it offers an instructive example of the function of logic in the philosophy of science. Witness, for instance, the esoteric-looking question of what the different readings of a question like (19) are and how they place different requirements on its (conclusive) answers. These questions turn out to be relevant to the issue of how we are to conceptualize the outcomes of controlled experiments and to suggest that we have a closer look at the role of these outcomes in actual scientific investigations.

NOTES

* The research reported here was made possible by NSF Grant #IST-8310936 (Information Science and Technology, Principal Investigators Jaakko Hintikka and C. J. B. Macmillan).

[1] See, *inter alia*, Jaakko Hintikka, *An Interrogative Model of Inquiry and Some of Its Applications* (forthcoming); 'On the Incommensurability of Theories' (forthcoming); 1985, 'A Spectrum of Logics of Questioning', *Philosophica* **35**, 135–50; 'The Logic of Science as a Model-Oriented Logic', in Peter Asquith and Philip Kitcher (eds.), *PSA 1984*, Philosophy of Science Association, East Lansing, MI, pp. 177–85; 1982, (with Merrill B. Hintikka) 'Sherlock Holmes Confronts Modern Logic: Towards a Theory of Information-Seeking By Questioning', in E. M. Barth and J. L. Martens (eds.), *Argumentation: Approaches to Theory Formation*, John Benjamins, Amsterdam, pp. 55–76.

[2] This is more than a matter of jargon, however. One of the main advantages of the interrogative model is that it enables us to study strategies of research and not just

one-time scientific inferences. Now game theory is the most important conceptual tool of strategy research; hence its concepts are extremely handy in studying the interrogative model.

3 The original presentation is E. W. Beth: 1955, 'Semantic Entailment and Formal Derivability', *Mededelingen van de Koninklijke Nederlandse Akademie van Wetenschappen, Afd. Letterkunde*, N. R. vol. 18, no. 13, Amsterdam. As a logical proof technique, the tableau method is not only equivalent but virtually identical with a Gentzen-type sequent method, except that the closed tableau which may result from the Beth method is in the Gentzen method read in the reverse order, i.e., from the bottom up. What the Beth method allows is to view the tableau construction as a frustrated attempt to bulid a countermodel in which the left-column entries are all true and right-column entries are false.

4 For the concept of presupposition (and other basic concepts in the theory of questions and answers), see Jaakko Hintikka: 1976, 'The Semantics of Questions and the Questions of Semantics', *Acta Philosophica Fennica* **28**, No. 4, Helsinki

5 Interesting discussions of the peculiarities of clinical inquiry are few and far apart. Typically, little attention is paid to the structural differences of clinical and pure research.

6 For syntax, see Chomsky's methodological writings, and for semantics, see Jerrold J. Katz: 1972, *Semantic Theory*, Harper & Row, New York.

7 *Critique of Pure Reason*, preface to the second edition, B xiii of the original.

8 The logic of such questions is extremely tricky, and left completely unexplained in the earlier literature on the logic and semantics of questions. For a brief discussion, see my papers, 'Questions With an Outside Quantifier' in R. Schneider, K. Tuite and R. Chametzky (eds.), *Papers From the Parasession on Nondeclaratives*, Chicago Linguistics Society, 1982, pp. 83–92; and 'On Games, Questions, and Strange Quantifiers', in Tom Pauli (ed.), *Philosophical Essays Dedicated to Lemart Åqvist*, Department of Philosophy and the Philosophical Society, Uppsala, 1982, pp. 159–69.

9 See, e.g., 'A Spectrum of Logics of Questioning', note 1 above.

10 For branching quantifiers, see my 1974 paper, 'Quantifier vs. Quantification Theory', *Linguistic Inquiry* **5**, 153–77, reprinted in Esa Saarinen (ed.), *Game-Theoretical Semantics*, D. Reidel, Dordrecht; 1979.

11 See note 10 above.

12 Forthcoming.

13 When N. R. Hanson introduced the ideas of theory-ladenness and concept-ladenness of observations, he was careful to make it clear that "here . . . the psychological is a symbol of the logical". (See N. R. Hanson: 1958 *Patterns of Discovery*, Cambridge University Press, Cambridge chap. 1, especially p. 17.) Yet in later discussion the problem has got thoroughly muddled as witnessed, e.g., by appeals to the psychology of perception against the possibility of a theory-neutral language. (Cf., e.g., T. S. Kuhn: 1962 *The Structure of Scientific Revolutions*, University of Chicago Press, Chicago, pp. 112–13.) Yet plainly, the question whether one's concepts and theories exert causal influence on what one perceives has no relevance whatsoever to the logical structure of the scientific process.

14 See, e.g., Max Jammer: 1966, *The Conceptual Development of Quantum Mechanics*, McGraw-Hill, New York, chap. 1.

[15] Oskar Becker: 1951, *Die Grösse und Grenze der mathematischen Denkweise*, Karl Alber, Freiburg, sect. 2.2.
[16] T. S. Kuhn: 1976, 'Mathematical versus Experimental Traditions in the Development of Physical Science', *The Journal of Interdisciplinary History* **7**, 1–31; reprinted in T. S. Kuhn, *The Essential Tension*, University of Chicago Press, Chicago, 1977, chap. 3.
[17] Cf. Jaakko Hintikka and Unto Remes: 1974, *The Method of Analysis*, D. Reidel, Dordrecht, chap. 9; and Jaakko Hintikka: 1978, 'A Discourse on Descartes, Method' in Michael Hooker (ed.), *Descartes: Critical and Interpretive Essays*, The Johns Hopkins University Press, Baltimore, pp. 74–88.
[18] This does mean, of course, that the outcome of a micro-level inquiry needs to be considered incorrigible in some absolute sense. All I need to do to handle this possibility is to allow uncertain answers by Nature in the macro-level game, which I want to do anyway. See Section 12.
[19] G. W. Leibniz: 1969, *Philosophical Papers and Letters*, Leroy E. Loemker (ed.), D. Reidel, Dordrecht, p. 188; cf. p. 283.
[20] Cf., e.g., Jammer, note 14, chap. 2.
[21] See note 8.

Department of Philosophy
Florida State University
Tallahassee, FL 32306-1054
U.S.A.

8

THE CONCEPT OF INDUCTION IN THE LIGHT OF THE INTERROGATIVE APPROACH TO INQUIRY

1. THE INTERROGATIVE MODEL

This paper is a part of a larger enterprise. In the last few years I have developed an essentially new approach to inquiry, prominently including scientific inquiry.[1] This approach can be given the form of a model (or, more appropriately, a multidimensional spectrum of models) of inquiry, which I propose to call *the interrogative model of inquiry*. It can be used to understand a wide variety of phenomena and to solve a variety of problems. In this paper I will study one particular concept that can be put into an interesting systematic and historical perspective by means of the interrogative model, namely, the concept of *induction*.[2] In the study of this notion the interrogative model of inquiry turns out to be especially useful. Among other conclusions, we shall find interesting reasons for assigning induction in the accepted sense of the word a rather lowly place on the map of scientific methodology. At the same time, the interrogative model helps us to uncover a historically earlier sense of induction which in our days is virtually forgotten and to assign to it an important role in the scientific process.

I will assume familiarity with the main ideas of the interrogative model. There is in fact relatively little to be familiar with, as the main ideas on which this model is based are extremely simple. The model can be described in game-theoretical terms.[3] The model takes the form of a game which an idealized scientist, called the *Inquirer*, plays against *Nature* on a fixed model (universe of discourse). This model or "world" in practice is usually our actual world or some part of it. (Such parts as can serve as models of theories are often called in physics independent or isolated systems.) The game starts from a theoretical premise T. The Inquirer is trying to derive a preset conclusion C from T. At each stage, the Inquirer has a choice between a *deductive move*, in which a logical conclusion is drawn from what the Inquirer has reached already, and an *interrogative move*, in which the Inquirer puts a question to Nature and registers the answer, when forthcoming, as an additional premise. Speaking of such questions is what presupposes that a model of the combined language of T and C is given to which the questions pertain. Nature's answers are assumed to be true in this model. In the applications contemplated here, "questions put to Nature" are typically intended to be observations and experiments.

In different varieties of the model, different assumptions are made as to what kinds of questions Nature can answer. One important dimension is represented by different restrictions on the logical complexity of the available answers, as measured by the number of quantifier kind changes (i.e., changes from an evidential to a universal quantifier or vice versa) in the quantifier prefix of the answer. Quantifier-free answers are called $A^0 = E^0$ answers. A^{n+1}-answers have a prefix of the form $(\forall x_1)(\forall x_2)\ldots(\forall x_k)$ + an E^n-prefix, and E^{n+1}-answers have a prefix of the form $(\exists x_1)(\exists x_2)\ldots(\exists x_k)$ + an A^n-prefix. A^2-prefixes will also be called AE-prefixes.

Restricting Nature's answers to the $A^0 = E^0$ case is tantamount to restricting answerable questions to yes-or-no questions concerning the truth or falsity of atomic propositions in M. The restriction that limits Nature's answers to this case I have called the *Atomistic Postulate*. At first sight, it seems to characterize the logic of empirical sciences. For Nature will not tell us in one fell swoop what happens always and everywhere. All that she will directly inform the Inquirer of is what happens in particular cases—in particular observations or measurements. And this restriction is precisely what the Atomistic Postulate is calculated to capture.

It can in fact be argued that practically all recent philosophy of science has been based on the Atomistic Postulate, tacitly if not explicitly.[4] Yet this postulate is unacceptable, for when a controlled experiment is construed as Nature's answer to the Inquirer's question, the logical complexity of the answers is at least of the AE variety. Hence the logic of experimental sciences (as distinguished from purely observational sciences) is an AE logic rather than one characterized by the Atomistic Postulate.[5] This result prompts a major reevaluation of contemporary philosophy of science. The present paper is a part of that reevaluation.

2. CONVENTIONAL INDUCTION PLAYS NO ROLE IN THE INTERROGATIVE MODEL

The first observation that can be made here is that induction in our accustomed twentieth-century sense play absolutely no role in the original interrogative model. In this sense, induction means, in the first place, inference from particular cases to general truths and, secondarily, inference from particular cases to other particular cases. I shall here disregard the latter aspect, for the following reason: If inferences from particulars to particulars satisfy certain conditions, the principles according to which they are made are logically equivalent to principles governing inferences from particulars to generalizations.[6] Hence it suffices for most purposes to consider only inferences from particular instances to general laws.

As the reader can ascertain, there is in the original interrogative model absolutely no slot for the kind of inference that induction is supposed to be. The only inferences that take place in an interrogative inquiry are deductive. The other moves (other than deductive ones) include in the first place interrogative ones. In them, no inferences are drawn. Instead, a question is addressed by the Inquirer to Nature and (depending on conditions that are defined when the interrogative model is further specified) answered by Nature. In some varieties of the model, the Inquirer can, instead of an interrogative or a deductive move, perform a definitory move (introduce a new concept by an explicit definition) or strengthen the conclusion to be proved (assertoric move). But in none of these extensions of the interrogative model is anything like an inductive move possible.

The original unreconstructed interrogative model of inquiry hence already leads to a remarkable result. It shows that it is possible to develop a rich and realistic model of at least some central aspects of the scientific enterprise without as much as mentioning induction. *A fortiori*, if the interrogative model should turn out to be, not only the truth and nothing but the truth, but the whole truth, Hume's problem would play no role whatsoever in a serious theory of the scientific method and of the scientific process.

What is especially interesting here is that the most important extensions of the original interrogative model also fail to vindicate the received concept of induction. I cannot examine these extensions in detail here. The main idea is nevertheless easy to appreciate.[7] Instead of assuming that the answers given by Nature are always true, it can be assumed that each such answer is true only with a certain probability. This does not yet specify a unique model, for it does not tell us how the probabilities of a true answer on different occasions depend on (or are independent of) one another. By choosing these probabilities in different ways we can adapt the interrogative model to different evidential situations. For instance, if Nature's answers to repetitions of one and the same question are independent of one another, the best way for the Inquirer to ascertain that Nature's answer is veridical may be to repeat the same question and hope for the same answer. This corresponds to a scientific situation in which an experiment or observation does not involve a systematic bias or other systematic mistakes.

It may happen, however, that an answer by Nature to a given question makes it likely that Nature should give the same answer to its repetitions. Then the best strategy for ascertaining Nature's veracity may very well be to try to derive the same conclusion by an altogether different line of reasoning. In this way, the famous old idea of the consilience of scientific inference can be explained and vindicated.[8]

In all these different situations, however, we are still dealing with deductive rather than inductive reasoning. More accurately, we are dealing with strict inferences from merely probable premises. The is diametrically opposite to typical cases of inductive inference, which are nonbinding inferences from (typically indubitable) premises. It is seen, not only that inductive inferences are not incorporated in the extended interrogative model, but that there is no place for them in the interrogative model or in any of its most natural extensions.

3. THE PROBLEM OF INDUCTION AND THE ATOMISTIC POSTULATE

One can say much more here, however. Indeed, one can put the entire concept of induction in a sharper historical relief.

The concept of induction goes back to Aristotle's idea of *epagoge*.[9] Indeed, the Latin term *inductio* was first introduced as a translation of *epagoge*. It is far from clear, however, that *epagoge* really is the same idea as our received concept of induction, and it will be argued below that it is not. In any case, the "problem of induction," by which everybody means the problem of *justifying* induction, was only thrust to the forefront of philosophical discussion almost two thousand years after Aristotle by David Hume.[10] Why the time lag? Were pre-Humean philosophers too confused or too naive to appreciate the importance of the problem of induction? I don't think that they were. It can be shown, if I am right, that the ascendancy of "Hume's Problem" of induction is part and parcel of the same problem situation as contemporary philosophers' virtually unanimous assumption of the Atomistic Postulate. In fact, this is a natural occasion to put the entire concept of induction into an overall historical perspective.

First, it can be seen where the idea of inductive inference as an essential ingredient of the scientific process comes from. It is one possible reaction to the problem situation created by the assumption of the Atomistic Postulate. If this postulate is adopted, then a scientific theory (e.g., the initial theoretical premise T) cannot itself be derived by means of the interrogative procedure without an equally strong or stronger theoretical premise. For no nontrivial general laws can be deduced from particular propositions, for instance, from Nature's answers to questions concerning particular cases or particular situations. Hence, the suggestion goes, the two kinds of steps of the interrogative procedure, questions and deductive inferences, have to be supplemented by a third kind of step. This step is calculated to lead us from Nature's particular answers to general truths. And it is this role of generalizing steps that inductive inferences are supposed to play.

At the moment I am not discussing the intrinsic merits or demerits of this idea. What is relevant here is a simple historical prediction (or retrodiction) that can be based on my diagnosis of the rationale of the general idea of inductive inference. What follows from my analysis is that, for those philosophers and scientists who did not adopt that Atomistic Postulate, our received idea of induction was not likely to play the same role as it does for us. It was not needed as a supplement to deductive inference. There did not exist for such philosophers any Humean problem of justifying induction, for induction in our sense could largely be dispensed with. If such thinkers used the idea of induction, it was in some different kind of role altogether.

In particular, it can be expected that philosophers and methodologists who do not believe in the Atomistic Postulate will not abide by the current idea of induction as an inference from particular cases to a generalization. For, in the face of the Atomistic Postulate, it was precisely this kind of generalization that induction was supposed to mediate. If you don't accept the postulate, you don't need induction for this purpose.

Likewise, it is clear that the central role of "the problem of induction" is due to philosophers' adoption of the Atomistic Postulate.[11] Hence this problem of "justifying induction" can be expected to occupy only those philosophers who have tacitly accepted the postulate. This observation helps in fact to explain why "Hume's Problem" became a central problem in philosophy when it did—and also how, as I will show below.

Now, as a matter of historical fact, the Atomistic Postulate was not adopted in the earlier tradition of philosophy and science from Aristotle to Newton (inclusive), though for different reasons in the case of different historical figures. In the Aristotelian tradition, even perception can give us forms, which are already by themselves general concepts. Moreover, their presence in the soul *ipso facto* implied according to Aristotle awareness of certain general laws, namely, those laws that specify what forms necessarily accompany the given ones.[12] (Thinking of a certain form is for Aristotle to have it realized in one's soul. Hence what necessarily accompanies this form is also automatically present in the soul, that is, is also necessarily thought of.)

This rejection of the Atomistic Postulate by Aristotle is seen in other ways, too. Another indication is the fact, brought out by G. E. L. Owen and others, that the *phainomena* and *endoxa* which a theory was supposed to account for according to Aristotle, were not all particular facts but could include general laws.[13]

In Newton, his reason for dispensing with the Atomistic Postulate is different. It is Newton's firm belief in the experimental method as being able to give the Inquirer general laws as answers to experimental "questions put to nature."[14] Hence induction in our contemporary sense was not needed by Newton.

In either type of case, induction can therefore be expected to amount to something essentially different from our post-Humean conception. A widespread failure to appreciate this historical fact has in my judgment seriously impaired philosophers' understanding of the early history of the concept of induction. I don't think that most philosophers or historians really have a realistic idea of what Aristotle or Newton meant by induction.

4. EXPERIMENTAL QUESTIONS AND THEIR PRESUPPOSITIONS

In spite of all this, the interrogative model nevertheless assigns to a somewhat different (but historically authentic) concept of induction an extremely interesting role. In order to see this, a starting point is offered by the logical situation outlined in my paper, "What Is the Logic of Experimental Inquiry?"[15] When carried further, this analysis naturally, not to say inevitably, leads to further insights. In the earlier discussion, the reader was probably bothered by the question as to how much new force answers to AE questions (and to more complicated questions) really give us—and what kind of information they do yield. For from the perspective used there it may legitimately look as if the presuppositions of the questions in question are so strong that they are unlikely even to have been established in a realistic situation of inquiry and so strong that answers to them will yield relatively little new information anyway. Witness, for example, the question whose presupposition is:

$$\left.\begin{matrix}(\forall x)\,(\exists y)\\[1ex] (\forall z)\,(\exists u)\end{matrix}\right> S[x, y, z, u] \tag{1}$$

The logic of such sentences alone is almost tantamount to the entire second-order logic.[16] Hence the step from the presupposition (1) of a question put to Nature to its answer

$$K\,(\forall x)\,(\forall z)\,S[x, f(x), z, g(z)] \tag{2}$$

or even to the desideratum

$$(\exists f)\,(\exists g)\,K\,(\forall x)\,(\forall z)\,S[x, f(x), z, g(z)] \tag{3}$$

appears not to add terribly much new force to what the Inquirer already has.

Moreover, it is clear that the force of the presupposition (1) is largely due to the fact that it codifies certain independencies between different experimental variables. Now we normally think that such independencies can only be established experimentally. This expectation seems to me justified. But how can any experiment establish such independencies if they must have been found and incorporated into the presupposition of a question before the experimental question is put to Nature and answered by her?

These are genuine problems, and they force us to have another look at the most basic ideas on which the interrogative model is based. More specifically, we must have some second thoughts about the role of presuppositions. Right from the beginning, I could have built the model in a slightly different way. I could have eliminated the role of the presupposition of a question altogether as a prerequisite to asking the questions. Of course, I must then allow the answerer to respond, even when the question does otherwise qualify as an answerable one, by rejecting its presupposition rather than by providing the Inquirer with an answer to it.[17]

This rule change seems to result in a better model of scientific inquiry than the previous one, at least in many applications. It goes some distance toward solving the problems just pointed out. However, a little bit more has to be said of the failures of the presupposition of a question in different cases. For a propositional question,

$$(S_1 \text{ or } S_2 \text{ or} \ldots \text{or } S_k)? \tag{4}$$

the failure of the presupposition

$$(S_1 \vee S_2 \vee \ldots \vee S_k) \tag{5}$$

makes the question completely otiose, and likewise for simple wh-questions. But consider a question whose presupposition is:

$$(\forall x)\,(\exists y)\,S[x, y] \tag{6}$$

Such a presupposition fails if for *some* values of x *no* y satisfies S[x, y]. Then the Inquirer ought not to ask which y does the job for those fruitless values of x. But there is no harm in asking the question for other values of x. And even more so if the presupposition (6) is true but has not been shown to be true for some values of x. This should not bar the Inquirer from raising the question.

In such cases, Nature's response to a question whose presupposition used to be (6) is naturally taken to be of the same form

$$(\forall x)\,S[x, f(x)] \tag{7}$$

as before, with f such as to sustain the existential generalization from

$$K\,(\forall x)\,S[x, f(x)] \tag{8}$$

to

$$(\exists g)\,K\,(\forall x)\,S[x, g(x)], \tag{9}$$

except that f can now be a partial function, that is, defined only for some values of x. This can be expressed in the usual notation by turning the reply (8) (minus "K") into a conditional one:

$$(\forall x)\,((x_1 < x < x_2) \supset S[x, f(x)] \tag{10}$$

Instead of just one interval, we can of course in principle have in (10) several different nonoverlapping intervals. This, then, is what is natural to think of as Nature's answer to an AE-type question. I shall assume in the following that this is what Nature's answers are like. Nature's reply may also include a rejection of the presupposition for certain values of x:

$$(\forall x)((x' < x < x'') \supset \sim(\exists y)S[x, y]) \qquad (11)$$

I shall ignore this possible component of Nature's response in what follows, however. It is the less important and less interesting part of the reply.

Thus it is seen that the natural way of defining the interrogative model is to allow the Inquirer to put AE questions to Nature even when the presupposition is not established. Nature's reply is not the negative of the (former) presupposition, even when it is false. Rather, it is a restricted functional dependence like (10).

5. INDUCTION AS THE TASK OF EXTENDING LIMITED GENERALIZATIONS

From what has been said, it is seen that there is an important dimension of the progress of scientific inquiry which we have not yet considered and which has not been considered very often in recent philosophy of science. It is not, strictly speaking, any longer an application of the interrogative model. However, its importance is in effect predicted by the interrogative model in the light of the remarks just made.

This dimension concerns the extension of a generalization like (10) to further values of x, that is, a step from (10) to

$$(\forall x)((x_1' < x < x_2') \supset S[x, f(x)]) \qquad (12)$$

where $x_1' < x_1$, $x_2 < x_2'$.

More generally, the Inquirer has a number of partial generalizations

$$(\forall x)((x_{i1} < x < x_{i2}) \supset S[x, f_i(x)]) \qquad (13)$$

$(i = 1, 2, \ldots, k)$, where the intervals $x_{i1} < x < x_{i2}$ are assumed not to overlap. These may have been established in part as (partial) answers to the same question, and in part as (partial) answers to different questions.

The dimension I have in mind is now twofold: (i) the Inquirer is trying to extend a generalization like (10) or (13) to longer intervals, perhaps to an unrestricted generalization; and (ii) the Inquirer is trying to unify the different partial dependences (laws, functions) $f_i(x)$ into one law, for example, into one and the same mathematical law.

This dual dimension is the focus of this paper. I have shown why it is important from the systematic viewpoint offered to us by the interrogative model. I will return to this systematic significance later in this paper. The main historical thesis I am proposing also concerns the same dimension. What this thesis says is that the generalization and reconciliation task this dimension deals with is what was earlier meant by induction.

6. NEWTONIAN INDUCTION

This historical thesis deserves a few comments and explanations. Indeed, Newton's idea of induction is seen from his famous methodological statement in the *Opticks*:

> As in mathematics, so in natural philosophy, the investigation of difficult things by the method of analysis ought ever to precede the method of composition. This analysis consists in making experiments and observations, and in drawing general conclusions from them by induction, and admitting no objections against the conclusions but such as are taken from experiment, or other certain truths. For hypotheses are not to be regarded in experimental philosophy. And although the arguing from experiments and observations by induction be no demonstration of general conclusions, yet it is the best way of arguing which the nature of things admits of, and may be looked upon as so much the stronger by how much the induction is more general. And if no exception occur from phenomena, the conclusion may be pronounced generally. But if at any time afterward any exception shall occur from experiments, it may then begin to be pronounced with such exceptions as occur. By this way of analysis we may proceed from compounds to ingredients and from motions to the forces producing them, and in general from effects to their causes and from particular causes to more general ones, till the argument end in the most general. This is the method of analysis; and the synthesis consists in assuming the causes discovered and established as principles, and by them explaining the phenomena proceeding from them and proving the explanations.[18]

Here we can see the kind of concept of induction I propose to call the Newtonian one at work. Induction in this sense of the word occupies a niche in the scientific procedure indicated by the analysis carried out earlier in this paper. Induction is, in the simplest case, the step from a limited generalization of the form (10) to a less restricted generalization (12). In the extreme (ideal) case, the entire restriction (antecedent) in (10) is eliminated. This optimal case is what Newton describes as saying: "And if no exception occur from phenomena, the conclusion may be pronounced generally." What distinguishes a Newtonian inductive step from inductive inference in the twentieth-century sense is that its starting point is not a number of particular propositions but propositions which, logically speaking, normally possess already limited generality. For, as I have argued elsewhere, experiments could according to Newton yield as their result a general law.[19] (This is among other things seen from Newton's practice of resorting to results of experiments in proving general theorems and solving general problems in his *Opticks*.) Induction thus increases the generality of the propositions the Inquirer has established, but in an entirely different sense from the twentieth-century idea of induction as an inference from the particular to the general. For in Newtonian induction, the increase in generality is one of degree, not of kind. Newton is in fact quite consistent in speaking of induction as a way of arguing from "observations and experiments." Given his conception of experiment, this means that the starting point of an inductive step can already possess (limited) generality.

It is not clear in what sense a Newtonian induction is an inference, either. In principle, the extension of the scope of a generalization can take place experimentally. This is not what Newton means, however. For he says that "although the arguing from experiments and observations by induction be no demonstration of general conclusions, yet it is the best way of arguing which the nature of things admits of." What Newton has in mind is a fact of scientific life familiar to all experimentalists. Even though a controlled experiment (or equivalent) can establish a generalization, the actual range over which the controlled variable can be varied is typically very narrow, indeed so narrow that the resulting restricted generalization is useless in studying other phenomena unless it can be further generalized by widening its scope (as is the step from (10) to (12) above), even before other experiments and observations have been made to help us extend it. This implies that induction in the sense of widening the scope of a generalization is a corrigible step of argument in a sense which, for example, Nature's limited-scope answer to an experimental question is not. I will return to this point later.

At the same time, the quotations from Newton show how tempting it might easily seem to be to assimilate Newtonian induction to our latter-day namesake conception. It is indeed possible that Newtonian induction was intended to comprise as a degenerate special case steps from particular measurement results to a generalization. But this temptation is nevertheless misleading. It has in some cases even misled translators. For instance, this is illustrated by the Motte-Cajori translation of *Principia* where one of the crucial sentences is made to disagree with our interpretation:

> In this [experimental] philosophy particular propositions are inferred from the phenomena, and afterwards rendered general by induction.[20]

But there is no "particular" in the original, and the netural term "inferred" is a bad substitute for Newton's explicit word for deduction:

> In hac philosophia propositiones deducuntur ex phaenomenis, & redduntur generales per inductionem.[21]

The correct translation thus goes somewhat as follows:

> In this philosophy, propositions are deduced from phenomena and rendered general by induction.

If anything, Newton's terminology supports my interpretation, for a "proposition" normally meant in an axiomatic context of his day a *general* proposition.

Earlier in this paper, we saw that for Aristotle, phenomena could be the source of general truths (or general beliefs) and not just particular ones. The same is now seen to be true of Newton. For instance, in speaking of his laws of motion (which of course are supposed to be general truths *par excellence*) Newton, on one occasion writes:

> These Principles are *deduced from Phaenomena* and made general by Induction.[22] [emphasis added]

Since Newton uses "deduce" in a fairly strict sense, this makes sense only if Newton's "Phaenomena" are already general. Similar evidence is easily found elsewhere. (Cf. also what was established above.)

Thus those few hardy souls who have maintained that Newton's methodology was an "inductivist" one may after all have something going for them. But, of course, Newton was speaking of "induction" in a sense different from theirs.

7. NEWTON AND MATHEMATICAL MODELING

There is yet another similarity between Newton and Aristotle which has not been utilized in the literature on Newton. It is the role of induction in establishing the first principles of science in Aristotle. The way this oversight has pointed historians' and philosophers' attention in a wrong direction is seen in the views of those commentators who construe Newton's method as a construction of mathematical models.[23] This idea has been encouraged by Newton's relatively sparse mention of experiments and observations in the early books of the *Principia*. This seems to differentiate his procedure from that prescribed by the interrogative model, which apparently squares better with what Newton does in the *Opticks*.

The answer is that Newton assumes that most of the pertinent questions had been put to Nature before the definition and axioms (laws of motions) are formulated. We have in fact seen that Newton claimed in so many words that those laws (principles) were "deduced from Phaenomena & made general by Induction."[24] Hence the principles of the *Principia* do not just define a mathematical model. They are themselves not only somehow inspired by experience, as in any old mathematical model, but in fact derived from phenomena, Newton claims. This distinguishes Newton's procedure from typical hypothetico-deductive model construction, and it is well in keeping with the Aristotelian conception of a particular empirical science.

That Newton himself looked upon the *Principia* in this way is amply demonstrated by evidence. The following is a part of Newton's self-description of what he did in the *Principia*:

> The Propositions in the following book were invented by Analysis. But considering that the Ancients (so far as I can find) admitted nothing into Geometry before it was demonstrated by Composition I composed what I invented by Analysis to make it Geometrically authentic & fit for the publick. And this is the reason why this Book was written in words at length after the manner of the Ancients without Analytical calculations. But if any man who understands Analysis will reduce the Demonstrations of the Propositions from their composition back into Analysis (which is very easy to be done,) he will see by [what] method of Analysis they were invented. [And] by this means the Marquess de l'Hospital was able to affirm that this this [sic] Book was [presque tout de ce Calcul.] almost wholly of the infinitesimal Analysis.[25]

And here is another Newtonian self-description, this time anonymous:

> By the help of the new Analysis Mr. Newton found out most of the Propositions in his *Principia Philosophiae*: but because the Ancients for making things certain admitted nothing into Geometry before it was demonstrated synthetically, he demonstrated the Propositions synthetically, that the Systeme of the Heavens might be founded upon good Geometry. And this makes it now difficult for unskillful Men to see the *Analysis* by which those Propositions were found out.[26]

8. INDUCTION AND CONCEPT FORMATION

Already at the stage of analysis we have reached, a suggestive resemblance between Newtonian and Aristotelian induction is beginning to emerge. For Aristotle, the problem of induction was not first and foremost a problem of inference from particulars to a generalization. It was a problem of concept formation.[27] Particular cases were stepping-stones to the concepts or forms "induced" to be realized in the soul. Once they are formed, the laws governing them are obvious. (This is very closely related to Aristotle's rejection of the Atomistic Postulate.) Hence there is no such problem as the justification of induction for Aristotle. For Newton, too, the crucial step was the formation of the mathematical law on the basis of an experiment. The extension of this law to other cases is a lesser problem, and a problem that is at least partly mathematical (and hence conceptual) in nature, as we shall see later.

This similarity is not accidental, and it can be deepened by noting other points of contact between Newton and Aristotle.

9. HUME MISINTERPRETED NEWTON

Some light on the history of the problem of induction in thrown by its origin. It was made a centerpiece of epistemology and philosophy of science by Hume. Now Hume's approach to methodological matters is characterized concisely as a misinterpretation of Newton's methodology. Hume wanted to introduce what he called the experimental method into philosophy.[28] For Hume, however, the term 'experimental' was merely a near-synonym of 'empirical', as one can easily show.[29] In general, the good David did not have any idea of a genuine controlled experiment. In particular, Hume totally failed to understand Newton's idea that an experiment can yield general laws (dependencies). Instead, he firmly assumed the Atomistic Postulate. Hence the logico-methodological reasons which I showed to underlie the modern (post-Humean) preoccupation of philosphers with "Hume's Problem" are exemplified in an actual historical setting by the originator of this alleged problem himself.

I will return later to the relation of Hume's problem to Newton's explicitly acknowledged admission of the corrigibility of induction.

10. RECONCILING DIFFERENT PARTIAL GENERALIZATIONS

All this nevertheless amounts to treating only the simplest kind of inductive task, namely, the task of extending a generalization established for one interval of values of the controlled variable. (Cf. (1) above.) In the general case, the Inquirer faces the task of not only extending one interval but combining different kinds of functional dependence found to obtain over different (nonoverlapping) intervals. In this case, the problem is not simply extending a generalization but also reconciling with other partial generalizations. I will refer to this second aspect as the reconciliation problem.

This kind of reconciliation problem is not atypical in the history of science. A reader may in fact have here a déjà vu experience. For consider the problem situation which Planck faced and which I used as an example in the paper, "What Is the Logic of Experimental Inquiry?"[30] This problem situation is precisely one of reconciling partial generalizations with one another. Another important case in point is perhaps Einstein's discovery of the special theory of relativity, which may be considered a grand reconciliation of Newtonian mechanics and Maxwell's theory of electromagnetism. Other examples are easily found in the actual history of science.[31]

In Newton's own case, the combination and reconciliation problem is perhaps instantiated by his discovery of the law of gravitation. Newton had several ranges of two-body systems to generalize from, including those formed by falling terrestrial bodies and the earth; by the earth and the moon; and by the sun plus a planet obeying Kepler's laws. In this case, the reconciliation consisted mainly in ascertaining that the same gravitational constant was obtained in all the cases, for the laws themselves were of the same inverse-square form.

This kind of reconciliation task is an important variety of induction in the Newtonian sense of the word. It represents a problem situation which has played a major role in the development of science and which is therefore worth a study in its own right.

11. THE RECONCILIATION PROBLEM AND ARISTOTELIAN INDUCTION

Before studying the structure of the reconciliation problem, it is in order to point out a remarkable similarity between an inductive reconciliation of different restricted mathematical laws and Aristotelian induction. I have analyzed in detail Aristotle's conception of *epagoge* in an earlier paper.[32] Since induction was calculated to establish at least some of the first premises of a science according to Aristotle, and since those first premises were identified by Aristotle with definitions, the process of induction in Aristotle must be the same as a search for a definition. Now Aristotle gives a vivid example of the search for a definition in *Posterior Analytics* B 13. The notion to be defined there is *megalopsychia*, an interesting notion that is likely to intrigue any student of Greek moral philosophy in its own right. Now how is it that Aristotle proposes to look for a definition of *megalopsychia*? Here is his description:

> I mean, e.g. if we were to seek what *megalopsychia* is we should inquire, in the case of some *megalopsychia* we know, what one thing they all have as such. For instance, if Alcibiades is *megalopsychos*, or Achilles and Ajax are *megalopsychoi*, what one thing do they all have? Intolerance of insults; for one made war, one waxed wroth, and the other killed himself [because of this intolerance]. Again in the case of others, e.g. Lysander and Socrates. Well, if here it [sc. the common characteristic that makes them *megalopsychoi*] is being indifferent to good and bad fortune, I take these two things and inquire what both indifference to fortune and not brooking dishonour have that is the same. . . . then we must again inquire if what we have now got have anything that is the same (with still other cases)— until we come to a single account; for this will be the definition of the object.[33]

The analogy with the reconciliation problem is striking. What Aristotle is in effect saying is that one first has to establish a number of restricted generalizations, each specifying the defining characteristic of *megalopsychia* in one type of case. The task of finding the definition of *megalopsychia* then consists in reconciling these partial generalizations, that is, in finding what (if anything) the different partial definitions have in common. In my paper "Aristotelian Induction" I showed that the structure of this kind of definition seeking is precisely the same as the so-called syllogistic induction Aristotle describes in *Prior Analytics* B 23.[34]

The differences between a typical Aristotelian induction and the inductive reconciliation problem are also clear. First, for Aristotle the partial generalizations to be integrated with one another are not established experimentally, but by examining our own reasons for calling a class of people *megalopsychoi*. Second, the partial generalizations in question are typically qualitative and not quantitative. However, neither of these differences affects essentially the similarity of the two reconciliation problems.

Here we have an interesting glimpse of what the cash value of the term 'induction' (or *epagoge*) really was in pre-Humean philosophy of science.

It is especially interesting historically to see that the Aristotelians of the early modern period explicitly connected Aristotle's *epagoge* with his description of the search of a definition in *Posterior Analytics* B 13 and otherwise interpreted *epagoge* along the same lines I have done. It is also known that the young Newton read the *Organon* as well as the Aristotelian methodologists of his day. What historical insights a study of his notebooks might yield remains to be seen.

12. THE STRUCTURE OF A RECONCILIATION PROBLEM

It is now in order to discuss the reconciliation problem in its own right. What is the structure of induction in the sense of reconciliation of different partial generalizations? If a simple answer were forthcoming, we would understand the nature of scientific discovery much better than we actually do. Certain things can nevertheless be said. The task is to transform the different mathematical laws to a form where they can be seen to be special cases of a more general regularity. For instance, what Planck did was to argue that the Rayleigh-Jeans law amounts essentially to

$$\frac{\partial^2 S}{\partial U^2} = \frac{\text{Constant}}{U^2} \tag{14}$$

where S is the entropy of an oscillator and U the energy.[35] In contrast, Wien's law goes together with the different law, namely:

$$\frac{\partial^2 S}{\partial U^2} = \frac{\text{Constant}}{U} \tag{15}$$

Planck reconciled the laws (14) and (15) by regarding them as special cases of:

$$\frac{\partial^2 S}{\partial U^2} = \frac{a}{U(U+b)} \tag{16}$$

For small values of U this reduces to Wien's law and for large values of U essentially to the Rayleigh-Jeans law, which had been experimentally established by Rubens and Kurlbaum.

Once again, the final reconciliatory step appears both trivial and arbitrary. Jammer says that this step, "though mathematically a mere trifle, was one of the most significant and momentous contributions ever made in the history of physics."[36] The significance is of course not due to this one step but is in the line of thought that led Planck to a point where the competing partial generalizations are directly comparable—or, rather, directly reconcilable. Such a line of thought normally takes the form of a mathematical manipulation of the laws in question. But such a manipulation of mathematical formulas must be guided by an insight into the actual physical situation. For instance, we can ask: How did Planck get from the original experimentally established radiation laws to the simple equations (14) and (15), which are not directly verifiable? In order to reach these equations, Planck has among other things to introduce a new concept, the concept of entropy, which he shunned earlier but which he now found himself forced to use. And this reliance on the concept of entropy is not just a facet of clever manipulation of mathematical expressions. It reflects Planck's newfound respect for statistical concepts in the analysis of physical situations. Unmistakably, such a combination of mathematical and physical considerations has something of the character of conceptual analysis. Small wonder, therefore, that Planck's line of thought led to a momentous new concept, namely, that of a quantum of energy.

Perhaps it is not too far-fetched to say that a physicist has to have some grasp of why each of the limited generalizations to be reconciled holds in its range, just as Aristotle assumed that in our search for a definition of *megalopsychia* we have to know, for each of the restricted ranges of cases we regard as stepping-stones toward a definition, why members of that class are called *megalopsychoi*: Socrates and Lysander because of their indifference to good and bad fortune, Alcibiades and Ajax and their ilk because of their intolerance to insult, and so on.

These observations also put in an interesting perspective the use of mathematics in an experimental science like physics. They show that the role of mathematics is far subtler than is usually spelled out. Mathematics is not just an aid in registering the outcomes of experiments. It is a tool, albeit not a mechanically applicable tool, that can be used to *analyze* a physical phenomenon. Mastering the mathematics used in a physical theory is very closely related to mastering the basic physical concepts of this theory. (Just think of what a crucial breakthrough von Neumann's work on the mathematical principles of quantum theory was for the actual development of this branch of physics.) In particular, the experimentation with mathematical equations which is calculated to bring the different mathematical laws (empirical generalizations) under the same roof inevitably takes on something of the character of conceptual analysis of the relevant physical situation.

The subtlety of the reconciliation problem shows how far a cry actual scientific argumentation is from the inductivist paradigm of making generalizations from particular cases. By the same token, it shows how much interesting argumentation is left untreated by the hypothetico-deductive conception of science.

Admittedly, there is a certain similarity between a reconciliation problem and the problem of establishing an inductive generalization in the twentieth-century sense of the term. In both uses, an inquirer moves from special cases to a more general proposition. There nevertheless are important differences between the typical forms of modern induction and of the task of fitting several restricted laws under a more general one (in the sense of a law defined for a larger range of argument values). For one important thing, the restricted laws have to be reinterpreted in the reconciliation process. They cannot simply be considered instances of a generalization. For this reason, the reconciliation problem cannot, for example, be dealt with by means of existing inductive logics.

13. THE UNCERTAINTIES OF INDUCTION

Newton acknowledges that induction in his sense is not an incorrigible step:

> And although the arguing from experiments and observations by induction be no demonstration, yet it is the best way of arguing which the nature of things admits of.[37]

It is of interest to ask, however, what in practice is the source of the uncertainty of a Newtonian (step of) induction. Does the source of trouble perhaps lie in unexpected discontinuities of nature? An experiment can establish a precise mathematical function. If this function is analytic, it has a Taylor expansion. That means that its behavior for the entire range of values for which it is defined by the same Taylor series is determined by its behavior in an arbitrarily small neighborhood of any point for which it is defined. Hence the kind of extension that is involved in "extrapolating" an analytic function seems to be a purely mathematical operation, and not conjectural at all. The only possible source of trouble seems to be hidden discontinuities of the function or of its derivatives outside the range of cases in which it has experimentally been found to hold.

It turns out, however, that this is not the only or even the main source of the uncertainty of Newtonian induction. Consider, for instance, Planck's problem situation. The two radiation laws he was trying to reconcile of course could not be considered precise special cases of a more general law. Rather, they were good approximations to the general law for different ranges of values, so good, indeed, that approximation fell within the range of observational error.

Hence the logic of Newtonian induction appears to be inextricably intertwined with the notions of approximation and of the margin of experimental or observational error. This is not the case logically speaking, however. Logically, the problem of inductive reconciliation of partial laws can arise and have the same logical properties also when the approximative character of the partial laws plays no role. For instance, Newton knew that Kepler's laws could not be completely accurate, because of the gravitational influence of different planets on one another's movements. Likewise, the orbit of the moon around the earth was known by him to be subject of the sun's perturbatory influence. But the logical problem of reconciliation would have been the same even if he had in both cases been able to observe pure two-body systems.

In Planck's case, the two radiation laws are likewise approximations only. However, they would be precise laws if we could push the value of one of the variables to its limiting values (zero and infinity, respectively).

Hence what the crucial judgment-call typically is in a Newtonian induction is an independence or near-independence assumption. It has to be assumed either that the functional dependence that has been established is independent of further "hidden variables" or else that it is so nearly free from further factors that it can be considered the precise law for a suitable special case or range of values. Once again, the source of logical strength in scientific inference turns out to lie in establishing or assuming functional independence. This is in keeping with what was found in my paper "What Is the Logic of Experimental Inquiry?"[38]

Hence the logic of Newtonian induction is not essentially dependent on the quantitative character of the partial laws that have been established or on their character as approximations. In principle, a similar problem situation can arise in a theory expressible in a first-order language.

Notice also that this kind of uncertainty of Newtonian induction is rather different from the general uncertainty to which Hume called philosophers' attention.

Among other observations, we can now see that Popper's criticism of Newton's claim for having derived his law of gravitation from phenomena is without any force whatsoever.[39] Popper bases his criticism on the fact that Kepler's laws, which were the starting point of Newton's line of thought, cannot be strictly true if Newton's law of gravitation holds, for they do not allow for the gravitational interaction between different planets. How could Newton possibly have derived the true law of gravitation from false empirical generalizations, Popper asks? Yet we have seen that the merely approximate character of Kepler's laws is completely beside the point in evaluating the kind of "inductive" reasoning Newton used. It was perfectly possible for him to think of Kepler's laws as holding strictly for certain two-body systems. For those systems, the law of gravitation was, strictly speaking, implied by the laws of motion of a planet. The inductive step was then to note that the resulting law, established so far only for a certain range of two-body systems, agrees with other partial generalizations established for other ranges of cases. Popper's criticisms show merely that he has not begun to understand Newton's actual conception of scientific inference in general and of the role of induction in science in particular.

NOTES

1. See the following papers of mine: "Knowledge Representation and the Interrogative Approach to Inquiry," forthcoming in a survey volume edited by Keith Lehrer

and Marjorie A. Clay; "What Is the Logic of Experimental Inquiry?" *Synthese* 74 (1988): 133–190; "The Interrogative Approach to Inquiry and Probabilistic Inference," *Erkenntnis* 26 (1987): 429–442; "The Logic of Science as Model-oriented Logic," in *PSA 1984*, ed. P. Asquith and P. Kitcher (East Lansing, Mich.: Philosophy of Science of Association, 1984), 1: 177–185; "A Spectrum of Logics of Questioning," *Philosophica* 35 (1985): 135–150; "Questioning as a Philosophical Method," in *Principles of Philosophical Reasoning*, ed. James Fetzer (Totowa N.J.: Rowman and Allanheld, 1984), 25–43; "Rules, Strategies and Utilities in Interrogative Games," in *Cognitive Constraints on Communication*, ed. Lucia Vaina and Jaakko Hintikka (Dordrecht: D. Reidel, 1984), 277–294; (with Merill B. Hintikka), "Sherlock Holmes Confronts Modern Logic," in *Argumentation: Approaches to Theory Formation*, ed. E. M. Barth and J. L. Martens (Amsterdam: Benjamins, 1982), 55–76.

2. For a lucid and informative survey of the problem of induction in its twentieth-century sense see G. H. von Wright, *The Logical Problem of Induction*, 2d ed. (New York: Macmillan, 1957).

3. The point in using game-theoretical concepts and conceptualizations is of course not to be able to use the mathematical results of game theory, but to do justice to the importance of research strategies in science, modeled in the interrogative model by different questioning strageties.

4. Later (see sec. 9 below) it will be indicated how the Atomistic Postulate can serve to motivate an inductivist approach to the progress of science. By almost the same token, if a philosopher of science rejects the idea of inductive inference, virtually the only remaining alternative is to consider general scientific theories as hypothetical constructs in the sense that they are not derived from experience by any rule-governed procedure. Theories can only be tested against evidence via their observable consequences, but not derived from them. This is, roughly speaking, the hypothetico-deductive conception of science, which, like the inductivist one, has to be radically reconsidered if the Atomistic Postulate in given up.

Other views, including certain Kuhnian ones, are also arguably inspired in part by a tacit belief in the Atomistic Postulate.

5. This is argued in "What Is the Logic of Experimental Inquiry?" (n. 1 above).

6. In Carnap's terms, the characteristic function of an inductive method can be such that it assigns definite prior probabilities (between zero and one) to generalizations. Conversely, the prior probabilities of generalizations determine partly the characteristic functions (right betting ratios of the next randomly observed individual). Cf. here, e.g., Jaakko Hintikka and Ilkka Niiniluoto, "An Axiomatic Foundation for the Logic of Inductive Generalization," in *Studies in Inductive Logic and Probability*, ed. Richard Jeffrey (Berkeley, Los Angeles, London: University of California Press, 1980), 157–181.

7. Cf. here "The Interrogative Approach to Inquiry and Probabilistic Inference" (n. 1 above).

8. Ibid., especially pp. 432–435.

9. See here Jaakko Hintikka, "Aristotelian Induction," *Revue Internationale de Philosophie* 34 (1980): 422–439; K. von Fritz, *Epagoge bei Aristoteles*, Sitzungsberichte der Bayerischen Akademie der Wissenschaften, Phil.-hist. Klasse, Heft 3 (Munich: C. H. Beck, 1964); D. W. Hamlyn, "Aristotelian Epagoge," *Phronesis* 21 (1976): 167–184; Gerd Buchdahl, *Induction and Necessity in the Philosophy of Aristotle*, Aquinas Paper no. 40 (London: Aquin Press, 1963).

10. Cf. Jaakko Hintikka, "Concepts of Scientific Method from Aristotle to Newton," *Knowledge and the Sciences in Medieval Philosophy* (Proceedings of the Eighth International Congress of Medieval Philosophy, ed. Monika Asztalos et al.), *Acta Philosophica Fennica* 48 (1990): 72–84.

11. Notice, however, that the restrictive consequences of the Atomistic Postulate can be counteracted to some extent by strengthening the initial theoretical premise or premises T from which an interrogative inquiry starts. Hence induction became a problem for philosophers only after they accepted the Atomistic Postulate and rejected such sources of general truths as innate ideas. This is the underlying reason why the problem of induction became Hume's problem and, not, say, Duns Scotus's problem.

12. Cf. here Jaakko Hintikka, "Aristotle's Incontinent Logician," *Ajatus* 37 (1978): 48–65.

13. See G. E. L. Owen, "Tithenai ta phainomena," in *Aristotle et les problèmes de méthode*, ed. S. Mansion (Louvain: Publications universitaires, 1961), 83–103.

14. See here Jaakko Hintikka and James Garrison, "Newton's Methodology and the Interrogative Logic of Experimental Inquiry," forthcoming in the proceedings of the 1987 workshop in Israel on "Three Hundred Years of the *Principia*-Realism Then and Now."

15. See n. 1 above.

16. The two are equivalent in the sense that their decision problems have the same degree of difficulty. See Jaakko Hintikka, "Quantifiers vs. Quantification Theory," *Linguistic Inquiry* 5 (1974): 153–177.

17. Another way of handling the situation would be by means of the idea of partial answer.

18. I quote from the Dover Books edition (New York, 1952), 404.

19. Cf. here "Newton's Methodology and the Interrogative Logic of Experimental Inquiry" (n. 14 above).

20. (Berkeley and Los Angeles: University of California Press, 1934), 2: 547.

21. See the critical edition by Alexandre Koyre and I. Bernard Cohen (Cambridge, Mass.: Harvard University Press, 1972).

22. See I. Bernard Cohen, *Introduction to Newton's "Principia"* (Cambridge, Mass.: Harvard University Press, 1978), 241.

23. Cf., e.g., I. Bernard Cohen, *The Newtonian Revolution* (Cambridge: Cambridge University Press, 1980).

24. Cohen, *Introduction*, 241.

25. Quoted in Cohen, *Introduction*, 294.

26. Cohen, *Introduction*, 295.

27. For Aristotle, whatever necessarily accompanies a form in general accompanies it when it is realized in the human soul (cf. "Aristotle's Incontinent Logician" [n. 12 above]). Hence there is no difficulty according to Aristotle in seeing what is implied by a certain concept once this concept is fully formed in one's mind. Hence the primary premises of a science are for him definitions. (Cf., e.g., *Posterior Analytics* B 3, 90b24.) Hence also our problem of induction is replaced in Aristotle's thought by problems of definition seeking and concept formation.

28. See the subtitle to the *Treatise* (1739): *An attempt to introduce the experimental Method of Reasoning into Moral Subjects.*

29. Hume occasionally speaks of "millions of experiments" (see *Treatise*, Selby-

Bigge ed., 105) that are supposed to have confirmed his views. But they cannot conceivably be controlled experiments of the kind Newton habitually was referring to by the term 'experiment', and they could scarcely even all involve purposive interference with the normal course of events.

30. See n. 1 above.

31. Another example is the way in which van der Waals's gas law unifies Boyle's law with what is known experimentally of the behavior of gases at low temperatures (i.e., near their liquefaction point) and with the known behavior of liquids. This example is also instructive in that van der Waals's law is based on an analysis of the forces governing the interaction of molecules.

32. See n. 9 above.

33. I am using, with some modifications, Jonathan Barnes's translation: see *Aristotle's Posterior Analytics*, ed. Jonathan Barnes, Clarendon Aristotle series (Oxford: Clarendon Press, 1975), 74.

34. See n. 9 above.

35. For Planck's reconciliation of the two laws, cf. Thomas S. Kuhn, *Black-Body Theory and the Quantum Discontinuity 1894-1912* (Oxford: Clarendon Press, 1978); Max Jammer, *The Conceptual Development of Quantum Mechanics* (New York: McGraw-Hill, 1966); and Armin Hermann, *The Genesis of Quantum Theory (1899–1913)* (Cambridge, Mass.: MIT Press, 1971).

36. Jammer, *Conceptual Development*, 18.

37. Newton, *Opticks* (see n. 18 above).

38. See n. 1 above.

39. See Karl R. Popper, "The Aim of Science," *Ratio* 1 (1957): 24–35; reprinted in Popper's *Objective Knowledge* (Oxford: Clarendon Press, 1972), 191–205.

9

SEMANTICS AND PRAGMATICS FOR WHY-QUESTIONS

The importance of the study of why-questions should be obvious. An answer to a question of the form 'Why *X*?' is closely related to an explanation of the fact that *X*. Hence a satisfactory theory of why-questions can be expected to be the core of any satisfactory theory of explanation. Such a theory is a tall order, to judge from the frustrations of philosophers of science who have tried to develop one.[1] A case of point is the wealth of counterexamples and other criticisms that have been raised against Carl G. Hempel's[2] covering-law model of explanation, in spite of its being in many ways a natural and tempting one.

It turns out that the study of why-questions constitutes an instructive example of the methodology of logico-semantical theorizing in more than one respect. We shall outline a theory of the semantics and pragmatics of why-questions with three components. The first is a theory of questions and answers in general. Such a theory has been thought of as having been in existence for a long time already as a part of epistemic logic. It has turned out, however, that only the introduction of the notion of informational independence has made this treatment of questions and answers really general and brought out the greatest common denominator in different kinds of questions.[3] Accordingly, our first task is to sketch briefly the resulting theory, especially as a full-scale treatment of it is not available in the literature.

But the behavior of questions and answers cannot be understood fully in terms of what in the conventional terminology would be called their semantics. We also have to examine how questions and answers are used in information acquisition. Such a study would be conventionally classified as pragmatics. But what we need here is not pragmatics in what Yehoshua Bar-Hillel called the wastepaper basket sense. The relevant study of the use of language can be, and ought to be, as explicit and as systematic (as "formal," some people would misleadingly call it) as the study of the syntax or semantics of language. It can therefore admit of models as explicit as the formal languages of a logician are in the study of semantics. The theory of questions

[1] A recent volume in the philosophy of science is entitled *Inference, Explanation, and Other Frustrations*, John Earman, ed. (Berkeley: California UP, 1992).

[2] *Aspects of Scientific Explanation and Other Essays in the Philosophy of Science* (New York: Free Press, 1965), pt. IV.

[3] Hintikka, *The Principles of Mathematics Revisited* (New York: Cambridge, forthcoming), chs. 3–4.

and answers constitutes an interesting case study of pragmatics in this sense. In fact, the interrogative model of inquiry is precisely such a logical model of the use of questions and answers.[4] In brief, it is the central part of a genuine theory of the pragmatics of questions and answers. Hence it is another important item of our agenda here to sketch the fundamentals of the interrogative model of inquiry.

The explicitness and the use of logical tools in pragmatics (that is, in the study of the use of language) is not merely a matter of convenience or of the jargon one chooses to use. What there is in it for us is the possibility of bringing powerful results from the metatheory of logic to bear on the task at hand. Logical positivists are frequently criticized for an allegedly excessive use of logic in their theorizing in epistemology and in the philosophy of science. Our study of why-questions and of the covering-law model illustrates the fact that the opposite accusation would often be more appropriate. For it is precisely by using logical tools more powerful than the ones logical positivists had at their disposal that we can make essential progress here. These previously unused opportunities will be illustrated later in this paper.

I. TOWARD A THEORY OF WHY-QUESTIONS

But the study of why-questions brings in still further methodological viewpoints. The general theory of the semantics of questions and answers does not automatically accommodate why- and how-questions. The reason for this failure might be thought to be the greater complexity of the semantics of why-questions as compared with that of the more thoroughly (or perhaps more successfully) analyzed types of questions, such as who-, where-, and when-questions. The theory of such "normal" wh-questions is by this time firmly under control, unlike that of why-questions, and the explanation of this discrepancy is thought to lie in the complexity of the semantics of why-questions. Often, this complexity is thought to be related to the covering-law model of explanation. Thus, an answer to a why-question is supposed to involve an initial condition and a covering law, over and above the explanandum. Sometimes, the connection between the initial conditions and the explanandum is thought to be a modal one, with the initial condition necessitating (in some sense or another) the explanandum.[5]

[4] See, for example, Hintikka, "The Concept of Induction in the Light of the Interrogative Approach to Inquiry," in Earman, pp. 23–43.

[5] An example of a logical treatment of why-questions exemplifying these characteristics is Antti Koura, "An Approach to Why-questions," *Synthese*, LXXIV (1988): 191–206.

We shall argue that this view is diametrically wrong. From a purely semantical vantage point, why-questions are not more complicated cases of questions in general, but rather cases so simple that they might even be thought to be degenerate cases. Because of that very degeneracy, the pragmatic tasks they can serve will be different from those which other kinds of questions can serve. Hence their characteristic behavior—including the way our standard vocabulary of replies, answers, and so on is applied to them—will be different. And that difference, as compared with most other kinds of questions, must therefore be accounted for in pragmatic terms, such as the interrogative model of inquiry, rather than logical or semantical ones in the narrow sense of these words. Thus, the theory of why-questions offers an interesting example of the interplay of semantical and pragmatical theorizing.

We shall sketch a kind of "transcendental deduction" of a theory of why-questions. As was indicated above, the theoretical basis of this deduction consists of two theories. The first is a theory of the logic and semantics of questions and answers, developed by means of epistemic logic. The other deals with the pragmatics of questions and answers in the sense of their use in information acquisition. It is the theory of interrogative inquiry.

II. SEMANTICS FOR QUESTIONS AND ANSWERS

Neither the semantics of questions nor the interrogative model of inquiry can be explained here in full detail (and with full proofs). Fortunately, a few well-chosen examples will suffice to show what goes on in these two approaches and what consequences these two themselves have for why-questions.

The epistemic logic of questions and answers will be explained first. Most of the relevant concepts can be explained by reference to a couple of examples. A simple who-question might be, for example:

(1) Who murdered Roger Ackroyd?

The crucial thing about (1) semantically is its *desideratum*. The desideratum of a question is a description of the epistemic state of affairs that the questioner would like to have brought about (in the normal use of wh-questions). In the case of (1), its desideratum is:

(2) I know who murdered Roger Ackroyd.

in symbols:

(3) $(\exists x) KM(x,r)$

or, equivalently:

(4) $K(\exists x/K) M(x,r)$

where K = *I know that*, and where the slash '/' exempts a logically active constituent (in this case $(\exists x)$) from the scope of another (in this case K).

All desiderata of questions can be brought to a form KS, where S is in a negation normal form. When we are dealing with desiderata of questions, S contains expressions of the form $(\exists x/K)$ and/or $(\vee/K)$. They indicate what the new knowledge is that an answer to a question gives to the questioner. In the sequel, it will be assumed that a transformation to such a normal form has been done.

The *presupposition* of a question is obtained from its desideratum by omitting all the slashed conditions from the $\exists$s and $\vee$s. For instance, the presupposition of (1) is:

(5) $K(\exists x)M(x,r)$

Here, (5) says merely that I know that Roger Ackroyd was murdered (compare (2) with (5)).

In the interrogative model, the presupposition of a question must have been established before the inquirer is allowed to ask this question.

A *reply* to (1) is a statement of the form:

(6) $KM(b,r)$

where b is any singular term. A reply does not normally imply the desideratum, that is, it does not fully serve the purpose for which the question is normally asked. In brief, it does not qualify as a (conclusive) *answer* to the question. For this purpose, an additional premise, the *conclusiveness condition* is needed. In the case of (1), the conclusiveness condition is:

(7) $K(\exists x/K)(b = x)$

which intuitively speaking means that:

(8) I know who b is.

The outcome of a conclusive answer to a question is the extra information (logical force) of its desideratum as compared with the information conveyed by its presupposition. What was just seen is that this extra information has two components in typical cases, codified respectively by a reply to the question and by the corresponding conclusiveness condition. All this is eminently natural, and can easily be generalized to other cases, including more complex ones. In some of them, the slash notation is indispensable, even though it is not unavoidable in the case of (1) nor indeed in this paper in general.

The kind of question perhaps most directly relevant to the theme of this paper is an even simpler one, namely, a propositional question. Its desideratum has the form:

(9) $K(S_1\ (\vee/K)\ S_2\ (\vee/K)...(\vee/K)\ S_k)$

that is:

(10) $K(\vee_i/K)S_i$ $(i = 1,2,...,k)$

In a more familiar notation, (9)–(10) could be written as:

(11) $(KS_1 \vee KS_2 \vee ... \vee KS_k)$

(Such a distribution of *K*s over the disjunction $(\vee/K)$ is nevertheless not always possible with more complex questions.) An example of a question whose desideratum is (9)–(11) might be:

(12) Does Stig live in Sweden, Norway, or Denmark?

The parallelism between such questions and the simple wh-questions exemplified by (1) can be heightened by comparing (12) with the question:

(13) In which Scandinavian country does Stig live?

This question (13) is of course similar to (1), except that the values of the crucial variable constitute in the case of (13) a known finite set.

The presupposition of (9) is, according to what was said earlier:

(14) $K(S_1 \vee S_2 \vee ... \vee S_k)$

Modifying some earlier treatments slightly, it can be said that a reply to such a propositional question is of the form:

(15) KS_o

where the corresponding conclusiveness condition is:

(16) $K((S_o \supset S_1)\ (\vee/K)\ (S_o \supset S_2)\ (\vee/K)...(\vee/K)\ (S_o \supset S_k))$

that is:

(17) $K(\vee_i/K)(S_o \supset S_i)$

which can be written in a metatheoretical notation:

(18) $K(\exists i/K)(S_o \supset S_i)$

Here the parallelism between the desideratum (17) (or (18)) of a propositional question and the desideratum (7) of a simple wh-question is conspicuous.

All this is unproblematic. The treatment of propositional questions applies independently of *k*.

In general terms, it can be said that the desideratum of a question is of the form *KS*, where *S* is assumed to be in negation normal form, that is, in a form in which all negation signs are prefixed to atomic formulas. This *S* can be thought of as a first-order sentence, except that it contains one or more expressions of the form $(\exists x/K)$ or $(\vee/K)$.

The corresponding presupposition is obtained from a desideratum by omitting all the slashes $/K$. The notions of a reply and an answer can likewise be defined in general terms, even though we shall not try to do it here. Even more generally speaking, the slashed $/K$ is now seen to be the precise counterpart to the question ingredient of natural languages.

III. THE SPECIAL PLACE OF WHY-QUESTIONS

From this general theory of questions and answers, we obtain a basis for a theory of why-questions and their answers by a strategy that is familiar from many important scientific contexts, namely, by pushing an important parameter characteristic of the situation in question to its extreme value, often zero or infinity. When we do so, some of the familiar factors lose their significance, while others are thrown into a sharper relief. For instance, Galileo is supposed to have discovered his version of the law of inertia by letting the angle of inclination where a decelerating frictionless ball is rolling to diminish to zero.

Here the relevant extreme case is $k = 1$. It was noted that the logic and semantics of propositional questions is independent of the number *k* of alternatives offered. In the case $k = 1$, however, everything seems to be trivialized. Not only do propositional questions constitute a simple subclass of questions. Logically speaking, the case $k = 1$ is so simple as not to have any logical structure on which one could lay one's hands. Indeed, even the question element $(\vee/K)$ disappears from the desiderata of such "questions."

At first sight, the special case $k = 1$ might seem so degenerate as to be trivial, pragmatically as well as logically. When $k = 1$, the desideratum of the propositional question (9) becomes:

(19) KS_1

But this is also the presupposition of the same question. Hence the transition from the presupposition to the desideratum does not provide any new information or otherwise change the situation—or so it seems. Asking and answering such questions seems to be an instance of the kind of pointless circularity that Aristotle[6] branded in his list of fallacies as "babbling" (ἀδολεσχεῖν).

Yet what we have here in reality is a spectacular instance of a philosophical and more generally logico-analytical fallacy. Ludwig Wittgenstein once said that the main source of mistakes in philosophy is a one-sided diet: one feeds oneself only one kind of examples. This is the fallacy which underlies the objection just envisaged. But in the present case, we can be more specific. The fallacy that here threatens logical and semantical analysts is to consider only one kind of use of the concepts they are analyzing. In many similar cases, the differences between different uses are not conspicuous, and one does not have available an explicit pragmatic theory that could be used to acknowledge and to conceptualize the relevant differences.

IV. INTERROGATIVE MODEL OF INQUIRY

Fortunately, in the present case, the theory of questions offers us an especially instructive case study in that the main relevant uses of questions and answers have been examined (and are being examined) explicitly and in some detail. In particular, it has been shown what the main uses of questions are and how they differ from each other. Here, the interrogative model of inquiry mentioned earlier is especially useful.

So what is this model? Once again, it is impossible to give an exhaustive account of this approach to inquiry, reasoning, and argumentation within the confines of one paper. The relevant basic ideas are nevertheless easily explained. The original idea was to think of knowledge seeking as a question-answer sequence, interspersed by logical inferences from given initial premises and from results already obtained. As a book-keeping device, a modified variant of the Beth tableau method can be used. It can be described by means of the same terminology as Beth's original tableau method.

The tableau construction starts from the initial premise or premises on the top of the left column and the ultimate conclusion the inquirer wants to establish on the top of the right column. The logical inference steps are governed by a variant of the usual deductive rules for tableau construction, the main difference being that no traffic is allowed from the right column to the left one. The only real novelty is that whenever the presupposition of some question occurs in the left column, the inquirer is allowed to address the question to a source of information ("an oracle"). Whether the ora-

[6] *De Sophistici Elenchi* 173a32—*On Sophistical Refutations,* E.S. Forster, trans. (Cambridge: Harvard, 1955).

cle will in fact answer is determined separately as a part of the specification of the particular game in question. If an answer is forthcoming, it is entered into the left column. In the simplest version of the model, which can be thought of as a game (in the sense of game theory) between the inquirer and a single oracle, it is assumed that the initial premises and the oracle's answers are all true.

The gain in information in the transition from the presupposition of a question to its desideratum—and the precise nature of this new information—is reflected formally by the addition of the independence indicator $/K$ to the operative ingredients of the presupposition. These new slashes are indices of the new knowledge reached when we reach a conclusive answer to the question. This gain in information is illustrated, for example, by a step from (5) to (4).

In the simplest version of the interrogative model, the ultimate conclusion is an initially given proposition without slashes. Questions only serve the purpose of establishing (proving) this given conclusion. The model is not used to answer questions by means of interrogative inquiry, that is, to reach new, previously unknown results.

In the earliest formulations of the interrogative model in the literature, the epistemic element was hidden, and the language used was merely an ordinary first-order language. It turned out to be impossible to capture the process of answering questions by means of interrogative inquiry with such an approach. In order to do so, the epistemic elements $(\exists x/K)$ and/or $(\vee/K)$ in the ultimate conclusion have to be explicitly represented. That cannot be done without the epistemic notation. In other words, the original nonepistemic model could not serve all the relevant purposes, nor indeed some of the most important ones. One of them is to show just how questions can be answered by means of a step-by-step inquiry. Such an explanation presupposes several things. Among them there are the following.

(A) It has to be recognized that questions play two fundamentally different roles in interrogative inquiry. The aim of the entire game may be to answer a question. Such a "big" question is called the *principal question* of the inquiry. But a part of the means of answering the principal question is to put "small" questions to the oracle. They are called *operative questions.* They have to be distinguished sharply from the principal question. A failure to do so is what was originally meant by the fallacy of *petitio principii.* It literally meant to ask ("petition") the principal question instead of an operative one.[7] We have

[7] See, for example, Hintikka, "The Fallacy of Fallacies," *Argumentation,* I (1987): 221–38; and Richard Robinson, "Begging the Question, 1971," *Analysis,* XXXI, 4 (1971): 113–17.

seen why this is a mistake. The inquirer is not supposed to put the principal question to any oracle but to answer it herself indirectly by putting several "small" or operative questions to the oracle.

This distinction between the two roles of questions is the kind of pragmatic distinction (difference in use) to which philosophers and linguists should pay especially keen attention, in order to avoid a one-sided diet of examples and "intuitions" which seem general but which on a closer scrutiny turn out to pertain only to one of the two functions of questions in inquiry.

(B) In order to indicate what the principal question is that is being answered through the entire line of operative questions, we need to express explicitly the epistemic element that was only tacitly presented in the original simple version of the interrogative model. Once we have the notation of the epistemic logic available to us which was explained earlier, we can indicate the question that the entire inquiry is calculated to answer simply by using the desideratum of this question as the ultimate conclusion of the interrogative tableau. For this purpose, we need the notation of epistemic logic. Of course, this notation must then be used throughout the inquiry. Basically, we have to think of all the sentences of the original simple tableau as being prefixed by K and the slash notation has to be used to express the different kinds of knowledge which the inquirer can bring to bear on his task and which figure in the relationships between presuppositions, replies, and answers.

How this is done in detail need not be spelled out here. What is crucial is that it can be done, relying on the epistemic logic mentioned earlier in this paper. Then the class of interrogative arguments that could be represented without the epistemic element becomes a subclass of all possible interrogative arguments. This subclass is characterized by the fact that no epistemic elements $(\exists/K)$ and/or $(\vee/K)$ occur in any answer to an operative question which do not already occur in its presupposition. A moment's thought shows that this is in effect the case of propositional questions with $k = 1$. (Notice that in the desiderata of such questions there literally is no queried ingredient $(\exists/K)$ or $(\vee/K)$.) In general, an entire inquiry can be handled without making the epistemic element explicit if in its ultimate conclusion there are no queried ingredients indicated by $(\exists/K)$ or $(\vee/K)$.

Formally speaking, the main peculiarity of the critical case $k = 1$ is that the desideratum does not contain any new ingredients of the form $(\exists/K)$ or $(\vee/K)$, that is, any queried element. This logical peculiarity of why-questions is closely related to the grammatical peculiarities of why-questions pointed out by Sylvain Bromberger;[8] for instance, the absence of any queried element in the logical counterpart to the grammatical fact that the *why* of an ordinary-language why-question is not correlated with any trace in the body of the question. For in other wh-questions, the wh-word represents, roughly speaking, the queried element that in the syntactical generation of the question is moved to the beginning of a question from its original position in the clause, leaving a trace where it originally was. In the case corresponding to $k = 1$, there is no trace just because there was no queried element to begin with.

V. WHY-QUESTIONS IN THE CONTEXT OF INQUIRY

Now we are approaching the crucial insight of this paper. In it the distinction between principal and operative questions plays a decisive role. Why was it that propositional questions (9) with $k = 1$ were thought of as being utterly devoid of any significance? Because their desiderata coincided with their presuppositions. Now operative questions are used, as was explained, precisely to move from the presupposition of a question to its desideratum, or, strictly speaking, to an answer that entails its desideratum. Answers to why-questions cannot do this. But all that this shows is that (and why) *why-questions cannot serve as operative* (small) *questions in inquiry*. There is nothing, however, which prevents a why-question from serving as *the principal question of an inquiry*. For from the interrogative model it is seen that the presupposition of the principal question of inquiry plays no role in the inquiry. It does not have to be established prior to the inquiry. An inquiry itself is not a movement from the presupposition of the principal question to its desideratum, even though each particular interrogative step in inquiry is a step from the presupposition of an operative question to its desideratum.

Thus, the case where the principal question of an entire inquiry is a degenerate propositional question with $k = 1$ has nothing absurd or even strange about it. It is merely the case in which the ultimate conclusion of the inquiry contains no question element $/K$. This case is not even new in the literature. In fact, in the simplest version of the interrogative model,[9] the presumed purpose of the inquiry was to establish a predetermined conclusion which was specified without any reference to epistemic considerations. It could have been asked, and occasionally was asked, what realistic purpose

[8] "What We Don't Know When We Don't Know Why?" in his *On What We Know We Don't Know: Explanation, Theory, Linguistics, and How Questions Shape Them* (Chicago: University Press, 1992).

[9] Cf. footnote 4.

such an enterprise could possibly serve. Since all answers are assumed to be true (and known to be true) in that simple version, and since the conclusion is already known, there does not seen to be any purpose of conducting an inquiry to establish a known result. There would be a point in doing so if the oracle's answers (and hence the inquirer's conclusion) are uncertain, for uncertain conclusions could be strengthened by deriving them in a new way. But that is just not what is assumed in the simplest form of interrogative inquiry.

Now we see one possible interpretation of such an interrogative inquiry according to the original narrow model. Those interrogative inquiries might prima facie seem pointless. What interest can there be in establishing a conclusion by means of an interrogative inquiry that was "known" (in the sense of being fixed and given) at the onset of the inquiry? What has been said provides an interesting partial answer. One use of inquiries with a given conclusion is that they can often serve to answer a why-question, that is, to figure out *why* the known ultimate conclusion is true.

Thus, we have found a slot for why-questions within the general theory of questions and answers outlined above. Why-questions can serve as principal questions of an entire interrogative inquiry.

It may be pointed out here that, from the fact that why-questions can serve only as principal questions in inquiry, it does not even follow that they cannot occur (with their answers) as steps in an interrogative inquiry. Clearly, they do so occur in the normal practice of reasoning and argumentation. What follows is merely that those why-questions do not operate like ordinary operative questions. They are principal questions in a lower-level inquiry, each question as the principal question of its own mini-inquiry. This is possible because interrogative inquiry can be, and frequently is, a many-leveled process.[10] This many-leveled character of inquiry is not an ad hoc postulation, but is rooted in the distinction between the two roles of questions in inquiry.

What is peculiar to the use of why-questions as operative questions of a higher-level interrogative inquiry is that their answers (that is, the outcomes of the lower-level processes of answering them) that are entered into the higher-order questioning process as new premises are not the ultimate conclusions of the lower-level interrogative inquiry. The reason is that those ultimate conclusions are, in the case of why-questions, known already at the outset of the inquiry.

[10] See here Hintikka, "Theory-ladenness of Observations as a Test Case of Kuhn's Approach to Scientific Inquiry," in David Hull et alia, eds., *Philosophy of Science Association 1992* (East Lansing, MI: PSA, 1992), pp. 277–86.

In other words, *in ordinary usage the notion of answer is used in a logically different way when it is applied to why-questions from the way it applies to other kinds of questions.* (What that different sense is will be explained later in this paper.) Roughly speaking, it can be said that the answer to a why-question is the explanation of the ultimate conclusion rather than this ultimate conclusion itself.

VI. ANSWERS TO WHY-QUESTIONS

This leads us to the next set of questions. Even though we have found a niche for why-questions within our general theory of questions and answers, we have not yet seen how the main concepts of our theory apply insofar as they apply to why-questions.

The notion of presupposition is the easy one. It was seen that, in the case of why-questions, the presupposition of a question is identical with its "answer" in the sense of the ultimate conclusion. This makes perfect sense. One cannot say that one is explaining why it is the case that *S* unless one knows that *S*. Hence we can speak of the presuppositions of why-questions in the same sense as in our general theory.

In contrast, the applicability of the notion of answer to why-questions requires further analysis. The precise notion of answer used in our general theory is an explication of the pretheoretical concept of answer. This pretheoretical concept can be characterized roughly by saying that an answer is what is sought when an inquirer is responding to a question. But if so, the pretheoretical notion of answer must mean different things in connection with why-questions and ordinary wh-questions. The pragmatic purpose of asking a why-question is different from that of asking, say, an ordinary wh-question. We have seen that what an inquirer is looking for in trying to answer other questions is the ultimate conclusion. In contrast, when an inquirer is trying to cope with a why-question, she is looking for the argumentative bridge between initial assumptions and the given ultimate conclusion. This bridge is "the answer" in that it is what is sought. Accordingly, we shall not call the ultimate conclusion of an inquiry which is calculated to answer a why-question an *answer*. Instead, we shall call it the *explanandum*.

But this leads to a further conceptual problem. Clearly, there has to be something more manageable than the entire inquiry process that could qualify as an answer to a why-question. Marshall McLuhan is wrong when it comes to why-questions. The argumentative medium of answering them is not the message. Somehow, this process must be summarized in a single sentence before we can speak of an answer.

This abstracting service is performed by answers to why-questions in an interesting way that can be spelled out by means of certain interesting metatheoretical results in logical theory. They throw interesting light on the pragmatics of why-questions. They also vindicate the connection between why-questions and the covering-law model of explanation which was foreshadowed in the beginning of this paper. Our next task is to explain these results.

VII. COVERING-LAW THEOREM

We turn our attention again to the pragmatics of why-questions. In raising a why-question, the inquirer is typically dealing with an explanandum that deals with circumstances that are not covered directly by a given initial premise T. This idea can be explicated by assuming that the explanandum contains a constant b (in the general case, a number of nonlogical constants) that does not occur in the initial premise T. The why-question in question is assumed (for the sake of an example) to pertain to b, in other words, to be expressible by something like the question 'Why is b such-and-such?' (cf. below for different selections of the explanandum). As an illustrative special case, we shall first consider a special case in which the explanandum is of the simple form $P(b)$, where P is a one-place predicate. We assume also that A is the totality of answers that the inquirer can obtain from the oracle. The underlying logic is assumed to be ordinary first-order logic. Because of the compactness of this logic, we can deal with T and A as if they were single sentences. Using A, we can now note that whenever $P(b)$ is derivable from T interrogatively, the following logical consequence relation holds:

$$(T \& A) \vdash P(b)$$

Suppose that we assume this and that we also rule out various ways in which the situation might be trivialized. More explicitly, we assume the following:

(i) $(T \& A) \vdash P(b)$
(ii) not $T \vdash P(b)$
(iii) not $A \vdash P(b)$
(iv) b does not occur in T
(v) P does not occur in A

Of these, (ii) and (iii) rule out obviously degenerate cases. Likewise, (iv) rules out situations in which the explanandum is not really a new case in that b is already mentioned in the initial premise T of the explanation. The last assumption, (v), is not crucial, and it can be thought of as implementing a kind of rough division of labor between explanation and theorizing. If P could occur in A and not only in T, the questioning process where the conjuncts of A are the oracle's answers could serve the purpose, not of explaining why $P(b)$, but of first building up the general theory that is to serve as the basis of an explanation. Such an inquiry, or a segment of inquiry, is properly speaking a part of theory building rather than explanation. Condition (v) serves to rule out such a mixture of theorizing and explanation. Not unexpectedly, this assumption turns out to be less than completely indispensable.

If these conditions (i)–(v) are satisfied, we can prove the following. There exists a formula H or more fully:

(21) $H[b]$

such that:

(22) (a) $T \vdash (\forall x)(H[x] \supset P(x))$
(b) All the constants of H occur both in T and in A (that is, in the oracle's answers) except for b.
(c) $A \vdash H[b]$

If condition (v) is given up, we can add a fourth condition:

(22) (d) If P does not occur in A, it does not occur in $H[b]$.

This metatheorem will be called (a special case of) the covering-law theorem.[11] Basically, its proof follows easily from William

[11] Covering-law-theorem (special case):
Let T and A be sets of closed first-order formulas and P a one–place predicate constant. Assume the following:

(i) $(T \,\&\, A) \vdash P(b)$
(ii) not $T \vdash P(b)$
(iii) not $A \vdash P(b)$
(iv) b does not occur in T
(v) P does not occur in A

Then there is a formula $H = H[b]$ such that:

(a) $T \vdash (\forall x)(H[x] \supset P(x))$
(b) All the constants of H occur both in T and in A (that is, in the oracle's answers) except for b
(c) $A \vdash H[b]$

If condition (v) is given up, we can still obtain (a)–(c) and we can add a fourth condition:

(d) If P does not occur in A, it does not occur in $H[b]$.

Proof: T and A are sets of closed first-order formulas. In virtue of the compactness of first-order logic, they can be dealt with as if they were single sentences.

According to assumption (i) $(T \,\&\, A) \vdash P(b)$. Hence:

Craig's[12] interpolation theorem. It is one of the logical results which were not available to the logical positivists but which can be used to develop their essential ideas further.

Our result shows that in a sense why-questions, operating as principal questions of interrogative inquiry, do indeed have the kind of structure that some earlier theorists had tried to find in the semantics of why-questions. In particular, a why-question which operates as the principal question of an inquiry and which satisfies the exceedingly simple conditions (i)–(iv) inevitably involves an "initial condition" (21) and a covering law (22)(a).

A generalization of this result is obtained easily by letting the explanandum be an arbitrary sentence containing b, say, $C[x]$. Then everything goes through mutatis mutandis. The condition (d) will now say the following:

(d)* Those nonlogical constants of $C[b]$ which do not occur in A do not occur in $H[b]$.

The role of covering laws in explanation[13] throws light on the sense in which explanations are answers to why-questions—and vice versa. It is especially remarkable that the existence of the covering law:

$$(\forall x)(H[x] \supset P(x))$$

was obtained from assumptions in which nothing was assumed about the logical form of T or A and which merely ruled out (with the possible partial exception of (iv)) degenerate special cases.

(A) $A \vdash T \supset P(b)$

Looking at (A), we can see that from (iii) it follows that A is consistent and that from (ii) it follows that the conclusion of (A) is not a logical truth. Hence William Craig's interpolation theorem applies and yields a formula $I[b]$ such that:

(B) $A \vdash I[b]$
(C) $I[b] \vdash T \supset P(b)$

and that all the constants of $I[x]$ are shared by T and A. In particular, P does not occur in $I[b]$, since it does not occur in A.

Then from (C) it follows that:

(D) $T \vdash I[b] \supset P(b)$

Since b does not occur in T, it follows from (D) that:

(E) $T \vdash (\forall x)(I[x] \supset P(x))$

From (E) and (B) it is seen that $I[x]$ can serve as $H[x]$, q.e.d.

[12] "Three Uses of the Herbrand-Gentzen Theorem in Relating Model Theory and Proof Theory," *Journal of Symbolic Logic*, XXII (1957): 269–85.

[13] Cf. our "Toward a Theory of the Process of Explanation."

It may be noted here that from (i) and (iii) it follows that T cannot be logically (conceptually) true.

Thus, why-questions (at least such typical why-questions as satisfy conditions (i)–(v)) involve a great deal of interesting logical structure. This structure is essentially of the kind that earlier theorists tried to foist by fiat on the semantics of why-questions. We can see now from where this structure comes. It is not part and parcel of the logic and semantics of why-questions if they are considered in the same way as other questions. This logical structure, however, is a consequence of the pragmatic situation prevailing in a typical occasion on which why-questions are used.

This structure also shows how the usual concepts related to questions apply most naturally to why-questions. When other questions are answered through interrogative inquiry, the end of the inquiry is the desideratum of the principal question. In this sense, to answer a principal question is to establish its desideratum. In the case of why-questions, the desideratum is known at the start. The end of inquiry is to establish the bridge between the initial premises and the ultimate conclusion. The covering-law theorem shows that to establish this bridge is to find the formula $H[x]$ and the initial condition $H[b]$. Hence, finding the initial condition $H[b]$ is what is aimed at in answering a why-question. It can therefore be called an answer to the why-question. It is important to realize that this sense of 'answer' is only pragmatically the same as that of the same term when applied to other kinds of questions. Semantically and logically speaking, it is a different notion. Alternatively, the covering law $(\forall x)(H[x] \supset C[x])$ can be considered as (a part of) an answer. In either case, such "answers" are like the answers to other kinds of questions, for example, in that they can be used as premises of further inquiry.

It is more generally speaking important to realize that this sense of 'answer' is different from the sense of the same term when used in the theory of other kinds of questions. It has other logical properties than answers in the earlier sense of the term.

VIII. ANSWERS TO WHY-QUESTIONS AS ABSTRACTS OF AN INTERROGATIVE INQUIRY

It nevertheless turns out that the covering laws and their antecedents—abstracts of the answers to the given why-question—have remarkable properties of their own. These properties spell out the sense in which the instantiated antecedents of covering laws can be thought of as answers to why-questions. By so doing, they also serve to vindicate this usage of the term 'answer'. It can be shown that the covering law provides a kind of abbreviated overview of the entire interrogative inquiry with the desideratum of a why-question as its ultimate conclusion. What a successful inquiry gives us when the ultimate conclusion is the explanandum of a why-question is an entire line of argument leading to the explanandum. But such a line of argument is not what we normally hope to receive as an answer to a question, including a why-question. The covering-law theorem uncovers one particular proposition (the covering law) and one particular formula (the antecedent of the covering law) which play a prominent role in the argument. They can therefore be construed as "answers" to the why question explained above.

There is more to be said here. This is seen among other things, from the way in which the individual constants of the Hempelian condition H show which new individual constants were introduced in the course of the inquiry by the oracle's answers. Speaking more generally and more strongly, the character of answers to why-questions by interrogative inquiry as summaries of the structure of this inquiry has a neat logical counterpart. In the proof of the covering-law theorem, the Hempelian covering law H was obtained essentially as the Craigean interpolation formula corresponding to the proof of $C[b]$ from T together with nature's answers. (More accurately, and more interestingly, $H[b]$ is an interpolation formula between the generalized "initial conditions" $A[b]$ and the explanandum $C[b]$ as mediated by the antecedent of the conditional $T \supset C[b]$.) As shown in Hintikka and Antti Koura,[14] such interpolation formulas can be considered as summaries of the proof that gave rise to them.[15] This is another significant result in logical theory that contributes essentially to our understanding of the logical behavior of why-questions. According to this result, since $H[b]$ was obtained as an interpolation formula, *the structure of the formula $H[x]$ reflects the structure of the interrogative argument through which the explanation can be accomplished.* Thus, an analysis of the argument by means of which the covering-law theorem is proved likewise shows in greater detail the precise sense in which the condition (21) reflects the entire interrogative argument that shows the truth of $C[b]$.

As we might put the same point, the H-formulas that we have dubbed answers to why-questions are really summaries of the entire procedure of answering them. This fact is illustrated especially clearly by those answers to why-questions which summarize some person's line of reasoning that led him to act in a certain way.

[14] "An Effective Interpolation Theorem for First-order Logic" (forthcoming).

[15] This result presupposes merely that the proof in question is in a normal form (cut-free form) that satisfies the subformula property.

This character of "reply" (21) as a kind of precis of the entire interrogative argument from T to $C[b]$ explains its role as an "answer" to a why-question. Obviously, if you can summarize an entire inevitable blow-by-blow account of *how* something happened you have done enough to show *why* it happened.

Thus, an analysis of the logical and semantical behavior of why-questions can (and must) rely essentially on the theory of interrogative inquiry. The ultimate reason for this is, of course, that why-questions can operate only as principal questions of an entire inquiry. In the form of a slogan, it can thus be said that any further study of why-questions will have to involve, or perhaps even be part of, a study of the interrogative model.

It is worth noting that no assumptions have been made here as to the content or as to the logical form (or to the content) of the explanandum. It need not be the description of a particular past event. It can contain quantifiers and references to future as well as past times.

IX. CHOOSING THE QUERIED ELEMENT

Instead of focusing on the one given individual b mentioned in the explanandum (that is, in the ultimate conclusion of an interrogative inquiry), the inquirer can focus on another one, or on several, say, $b_1, b_2, \ldots$. In that case, the explanandum can be expressed as $C[b_1, b_2, \ldots]$. If none of the constants $b_1, b_2, \ldots$ occurs in T, the covering-law theorem applies and an "initial condition" establishes the existence of:

(23) $H[b_1, b_2, \ldots]$

and a "covering law":

(24) $(\forall x_1)(\forall x_2)(H[x_1, x_2, \ldots] \supset C[x_1, x_2, \ldots])$

with the same properties as before.

The choice of b (or $b_1, b_2, \ldots$) amounts to the choice of that element of the explanandum which is taken to be in need of an explanation. This choice is a pragmatic one. It makes a difference to condition (23) of the covering law. This means that a why-question can pertain to some particular individual in a stronger sense. As Bengt Hansson[16] seems to have been the first to emphasize, a question like:

(25) Why did John go to New York on Tuesday?

can be taken in several different senses, depending on which ingredient of (25) is being stressed; for example:

[16] In an unpublished work circulated in 1974; see also Bas C. van Fraassen, *The Scientific Image* (New York: Oxford, 1980).

(26) Why was it John who went to New York on Tuesday?
(27) Why was it New York to which John went on Tuesday?
(28) Why was it on Tuesday that John went to New York?

A small extension of the ideas (and the notation) of epistemic logic helps to spell out what is going on here. In the same way as the independence (slash) notation applies to other kinds of expressions, it also applies to individual constants. Thus, let us assume that the desideratum of (25), without any particular emphasis, is:

(29) $KG(j,n,t)$

Then desiderata of (26)–(28) can be represented as follows:

(30) $KG((j/K),n,t)$
(31) $KG(j,(n/K),t)$
(32) $KG(j,n,(t/K))$

These can also be written as follows:

(33) $K(\exists x/K)(x=j \,\&\, G(x,n,t))$
(34) $K(\exists x/K)(x=n \,\&\, G(j,x,t))$
(35) $K(\exists x/K)(x=t \,\&\, G(j,n,x))$

This brings out nicely a connection that there obtains between (26)–(28) and the corresponding who-, where-, and when-questions. Surely, answering (26) involves establishing, among other things, who it was that went to New York on Tuesday. There is a difference between why-questions and normal wh-questions, however, in that, in why-questions, unlike usual wh-questions, the queried element is not represented by a wh-word or by its trace. This is in keeping with Bromberger's observation mentioned above.

In other respects, too, a closer examination of (30)–(32) shows that they capture the intended meanings of desiderata (26)–(28).

The queried element can also be a general concept. For instance, if the explanandum is $P(b)$ and the queried element P rather than b (as in the earlier examples), then the covering law has the form:

(36) $(\forall X)(H[X] \supset X(b))$

X. THE NOMIC CHARACTER OF COVERING LAWS

One difference between our analysis and some of the earlier ones is that our covering laws:

(37) $(\forall x)(H[x] \supset C[x])$

do not involve any modal element. It is nevertheless easy to explain by reference to our theory why it has been thought that there is such an element of necessity in (37). The basis of an explanation lies in the fact that in a sense (37) is not the real covering law in the first place. For if you renew the proof of the existence of the covering law, you shall see that $C[b]$ and A may share a number of nonlogical constants that do not occur in T. These constants can occur also in $H[b]$. As an example, assume that these constants are $c_1, c_2, \ldots, c_k$. Then what has been called above the covering law is of the form:

(38) $(\forall x)(H[x, c_1, c_2, \ldots, c_k] \supset C[x, c_1, c_2, \ldots, c_k])$

But if (38) is logically implied by T, as it is in the conclusion of the covering-law theorem, then so is:

(39) $(\forall x)(\forall y_1)(\forall y_2)\ldots(\forall y_k)(H[x, y_1, y_2, \ldots, y_k] \supset C[x, y_1, y_2, \ldots, y_k])$

But obviously, (39) is here the "real" covering law, because (38) still contains references to particular individuals $c_1, c_2, \ldots, c_k$. Hence, the minimal covering law (38) has a greater generality than first meets the eye, but only potentially, so to speak. It has implicit generality with respect to all of its nonlogical constants that do not occur in T. What has happened in the earlier modal accounts is that this generality has been mistaken for necessity. We can say that covering laws have a nomic character, but not a modal one.

XI. HOW-QUESTIONS

It is of some interest to ask what happens if b is allowed to occur in the initial premise of the covering-law theorem. In that case, no covering law is obtainable in general. This case clearly corresponds much more closely to how-questions than to why-questions. It is, in fact, characteristic of how-questions, unlike why-questions, that their answers (in the pragmatic sense of 'answer') are not expected to be capable of being summed up by a single explanans. Instead, answers to how-questions are expected to amount to listing all the several steps that lead from the given initial conditions to the outcome to be accounted for. Furthermore, in how-questions, unlike why-questions, there is no counterpart to the choice of some particular ingredient of the ultimate conclusion as the element to be accounted for. All these things are consequences of the failure of the covering-law theorem on the weaker assumptions that allow b to occur in T.

Thus, not all questions whose desideratum logically speaking does not contain $/K$ are why-questions, nor are all interrogative inquiries whose ultimate conclusion does not contain any queried element attempts to answer a why-question. What is involved may be a how-question rather than a why-question. Such a how-question may be answered by the same interrogative argument as the corresponding why-question. What makes the difference is the relation of the conclusion (fact to be accounted for) or, more specifically, of the queried ingredient of the explanandum, to the initial premise *T* of the account to be given. If the queried ingredient is, say, an individual *b*, then we can obtain an answer to a why-question only if *b* does not occur in *T*. Otherwise, the best we can hope for is an answer to a how-question. Thus, the difference between the two types of questions is not only pragmatic, but also structural.

XII. SUMMING UP

It has turned out that why-questions can be handled perfectly well semantically and pragmatically by means of the interrogative model of inquiry and the semantics of normal questions that this model presupposes. The notions of answer and presupposition have to be reinterpreted, however. A (conclusive) answer is not any longer a reply that entails the desideratum of the corresponding question. An answer in the new nonstandard sense is the covering law and/or its initial condition, and the presupposition of a why-question in the new nonstandard sense is the given conclusion that the entire inquiry is supposed to establish.

It might thus be said that what happens in why-questions is that the entire machinery of interrogative inquiry is put to a new kind of use—"new" of course only in comparison with the types of questions studied earlier in the literature.

Why-questions are like other questions in that attempts to answer them can always be thought to be carried out by means of interrogative inquiry. Where they differ from other questions is that an answer to a why-question is not the conclusion of the entire interrogative argument, as it is in other cases. (Strictly speaking, of course, the conclusion is the desideratum of the given direct question.) In the case of why-questions, the sought-for answer in the pragmatic (rather than logical) sense is the initial condition of a covering law. Or, equivalently, the answer can be taken to be the entire covering law, for the initial condition and the covering law determine each other uniquely, given the explanandum.

This explains how answers to why-questions can contribute new premises to an interrogative inquiry, even though the conclusion of the interrogative argument which provides the answer to a why-question is known ahead of time. The new premise is not that conclusion, but the covering law and/or its initial condition. This observation thus also shows more fully what kind of lower-level inquiry it is that we are dealing with in answering a why-question.

Our results also serve to correct a widespread misconception concerning what answering a why-question (in effect, explaining) amounts to according to the covering-law idea. It is often thought and said that, according to the covering-law model, explanation is a purely deductive task, namely, the logical derivation of the explanandum from the covering law plus the initial condition. On the analysis presented here, an explanation (answering a why-question) consists of an interrogative derivation of the initial condition and of the specific covering law starting from some background assumptions *T*. These background assumptions need not in any realistic sense be assumed to be themselves "covering laws." Unlike *T*, the covering law is not given at the outset of the explanation, nor is the initial condition *H*. Rather, the task of explanation means a search for a derivation of the explanandum from *T* interrogatively. This normally involves putting questions to nature and is therefore an empirical rather than purely logical task. The answer to the why-question is then extracted from the interrogative argument as it is by means of Craig's interpolation theorem.

JAAKKO HINTIKKA

Boston University

ILPO HALONEN

University of Helsinki

10

THE VARIETIES OF INFORMATION AND SCIENTIFIC EXPLANATION

1. The importance of information

The concept of information seems to be strangely neglected by epistemologists and philosophers of language. In many everyday situations, knowledge and information are nearly exchangeable terms; yet for every score of books with the phrase "theory of knowledge" in their titles there scarcely exists a single paper by a logician or philosopher dealing with the theory of information.[1] Again, the information that a sentence yields or can yield might very well seem to an ordinary man much more important than the so-called meanings of the terms it contains, or even the meaning of the sentence itself. Yet, with but few exceptions, philosophers of language have not devoted more than a vanishingly small part of their efforts to the theory of information as compared with the theory of meaning. Why this should be so, I do not know. Perhaps the fact that mathematicians and communication theorists largely succeeded in appropriating the term "information" for their special purposes a couple of decades ago has something to do with this.[2] I also suspect that it is much harder to talk persuasive nonsense about the quantitative concept of information than of the qualitative notions of knowledge and meaning. Be this as it may, the neglect is a regrettable one. In this paper, I shall try to call philosophers' attention to a few possibilities of correcting it. I have already tried to do so in some earlier papers[3]; the present one is partly a sequel to them and partly a new enterprise.

---

[1] One of the most interesting exceptions is constituted by the studies of Ernest W. Adams and his students of the role of information in certain important methodological connections. See ADAMS [1966], HANNA [1966].

[2] See e.g. SHANNON and WEAVER [1949], KHINCHIN [1957], CHERRY [1957].

[3] HINTIKKA and PIETARINEN [1966] and HINTIKKA[1968].

2. Information as the elimination of uncertainty

The philosophical study of the concept of information was started by Carnap and Bar-Hillel soon after 1950.[4] It was called by them theory of semantic information, and distinguished from the statistical information theory of communication theorists. Similar ideas had been informally expounded by Sir Karl Popper already in the thirties.[5]

The basic idea of their approach may be said to be one particular way of explicating the general idea that information equals elimination of uncertainty. In order to measure this uncertainty, a distinction is made between the different logical possibilities that we can express in a language. The more of them a statement s admits of, the more probable it is in some "purely logical" sense of probability. The more of them a statement s excludes, the less uncertainty does it leave, and the more informative will it therefore be. The probability $p(s)$ and information $\inf(s)$ of a statement s are thus inversely related.

In some earlier papers, I have examined how the different cases might be distinguished from each other (and weighted) on which a purely logical measure of information might be based in certain simple languages.[6] These studies have strongly suggested to me that no absolute answer to this question is forthcoming. The weights that the different cases have to be given will depend on the order which we expect (or are entitled to expect) to obtain in the universe. Thus we are forced (it seems to me) from a purely logical conception of probability toward some form of Bayesianism.

The basic idea that underlies the concept of information nevertheless remains essentially the same in all cases. The information of s is the amount of uncertainty we are relieved of when we come to know that s is true.

3. Unexpectedness versus content

If this is all there is to be said here, any monotonically decreasing function of $p(s)$ could be thought of as a measure of the information of s. In order to pick out a particular measure, we must say a little more of how we expect the concept of information to behave vis-à-vis the concept of probability. Some important requirements of this kind are in fact quite straightforward. For instance, if it is required that $\inf(s)$ be additive with respect to proba-

---

[4] See CARNAP and BAR-HILLEL [1952], BAR-HILLEL and CARNAP [1953], KEMENY [1953].
[5] POPPER [1935].
[6] See HINTIKKA [1965], [1966].

bilistically independent conjuncts, i.e. that $\inf(s_1 \& s_2) = \inf(s_1) + \inf(s_2)$ if s_1 and s_2 are independent with respect to a probability-measure p, then the inverse relation must be expressed by the definition

$$\inf(s) = -\log \mathrm{p}(s) \tag{1}$$

provided that certain natural further assumptions are made.[7]

However, this is not the only way in which the concept of information or content can be defined, for there are other requirements which we can legitimately impose on this concept and which turn out to presuppose entirely different ways of defining information. In so far as it can be claimed that all these different requirements we can use here are part of our ordinary presystematic idea of information, the incompatibility of these requirements will show that this presystematic idea is intrinsically ambiguous.

Alternatives to (1) are in fact easy to come by. One very natural possibility here is to define the content cont(s) of s by the even simpler formula

$$\mathrm{cont}(s) = 1 - \mathrm{p}(s). \tag{2}$$

The justification of definition (2) will be indicated later.

It has been suggested that (1) is to be thought of as a measure of the *surprise value* or the *unexpectedness* of (the truth of) s, while (2) is a measure of the *substantive information* or *content* of s. This suggestion often helps to understand the difference between the two.[8]

The formal relation of inf(s) and cont(s) is in any case straightforward; it can be expressed e.g. by

$$\inf(s) = \log \frac{1}{1 - \mathrm{cont}(s)}. \tag{3}$$

4. Measures of relative information

In terms of inf and cont, we can also define relative measures of unexpectedness and of content:

$$\inf_{\mathrm{add}}(s|t) = \inf(s \& t) - \inf(t), \tag{4}$$

$$\mathrm{cont}_{\mathrm{add}}(s|t) = \mathrm{cont}(s \& t) - \mathrm{cont}(t). \tag{5}$$

---

[7] A somewhat loose sketch of an argument to this effect is given by Törnebohm [1964] pp. 49–50. Cf. Cox [1961] pp. 37–38.

[8] Cf. Bar-Hillel [1964] p. 307. The logical basis of this distinction is clear enough. Somewhat roughly speaking, it may be said that cont measures the absolute number of possibilities which an item of information enables us to disregard, whereas inf measures the relative number of such possibilities, with respect to the total number of possibilities not previously eliminated.

These relative measures tell us simply how much information s adds to the information t has already given to us. This explains the use of the subscript. We shall also often call $\inf_{\text{add}}$ and cont_{add} *incremental* information (or unexpectedness) and *incremental* content, respectively.

It is easily seen that both these measures of relative information can be given a very natural expression, though different in the two cases:

$$\inf_{\text{add}}(s|t) = -\log p(s|t), \tag{6}$$

$$\text{cont}_{\text{add}}(s|t) = \text{cont}(t \supset s). \tag{7}$$

Relative unexpectedness is thus related to relative probability in the same way as absolute unexpectedness to absolute (a priori) probability. The content of s relative to t is the content of the statement that if t is true, then so is s.[9]

This last fact suggests a way of justifying definition (2). It embodies a requirement which we might very well want to impose on our concept of (informational) content in any case.[10] But if we do so, we are readily led to some such relation as (2). In fact, if we simply define

$$p^*(s) = \text{cont}(\sim s)$$

then it follows easily from the requirement

$$\text{cont}(s \,\&\, t) - \text{cont}(t) = \text{cont}(t \supset s) \tag{8}$$

that p* satisfies some of the axioms of probability calculus. For instance, by substituting $\sim s$ and $\sim t$ for s and t in (8) we at once obtain

$$p^*(s \vee t) = p^*(s) + p^*(t)$$

for the case in which $(s \,\&\, t)$ is a contradiction, provided that logically equivalent statements are assumed to be intersubstitutable. This verifies the usual additivity requirement. By putting $s = t$ in (8) we likewise obtain

$$p^*(s) = 0 \quad \text{if} \quad s \quad \text{is a contradiction}.$$

If the normalizing conditions $\text{cont}(s) \geqslant 0$ and $\text{cont}(s \,\&\, \sim s) = 1$ are assumed, it follows that p* satisfies all the axioms of the probability calculus (with the possible exception of Kolmogorov's axiom of denumerable additivity).[11]

[9] See BAR-HILLEL [1964] chapter 15, for a number of further results concerning incremental content and information.

[10] One way of motivating (7) is as follows: the information that s adds to the information of t must surely be the information of the weakest statement which will, when added to t, logically imply s. This statement is $(t \supset s)$; hence the validity of (7).

[11] It has been suggested by Savage that this axiom is perhaps not entirely indispensable for all interesting purposes; see DUBINS and SAVAGE [1965] p. 10 (with references to similar opinions previously aired by De Finetti).

In terms of this measure, cont can then be defined as in (2). In this sense, then, does the assumption of (8) easily lead to a definition of form (2).[12]

There is another sense in which we can speak of relative information and relative content, however, besides the incremental sense.[13] In some ways, the motivation of this sense is even more obvious. Instead of asking what information or content s adds to t, we can consider the world on the assumption that t is true, and ask how information or content could be characterized for this limited purpose. The resulting senses of information and content might be called *conditional* information and *conditional* content, in short $\mathrm{inf}_{\mathrm{cond}}$ and $\mathrm{cont}_{\mathrm{cond}}$. The way to define them is obviously to substitute relative probability for absolute probability:

$$\mathrm{inf}_{\mathrm{cond}}(s|t) = -\log \mathrm{p}(s|t) \tag{9}$$

$$\mathrm{cont}_{\mathrm{cond}}(s|t) = 1 - \mathrm{p}(s|t). \tag{10}$$

From (6) it is seen that $\mathrm{inf}_{\mathrm{add}}(s|t) = \mathrm{inf}_{\mathrm{cond}}(s|t)$. We can thus omit the subscripts $_{\mathrm{add}}$ and $_{\mathrm{cond}}$ in the case of inf, and speak simply of relative information or unexpectedness. However, the difference between $\mathrm{cont}_{\mathrm{add}}$ and $\mathrm{cont}_{\mathrm{cond}}$ is vital.

The difference between the two might be illustrated by means of the following example. Let us assume that we are dealing with the kind of languages which Carnap typically considers and in which the probability of a statement is the sum of the probabilities of all the state-descriptions compatible with it. Then $\mathrm{cont}(s)$ is the sum of the probabilities of all the state-descriptions excluded by s, and $\mathrm{cont}_{\mathrm{add}}(s|t)$ is the sum of the probabilities of all the state-descriptions excluded by s but not by t. In contrast to this, $\mathrm{cont}_{\mathrm{cond}}(s|t)$ is the sum of the relative probabilities of all the state-descriptions excluded by s relative to t.

5. Information relative to a subject matter. Transmitted information

An important further ambiguity in the notion of information can be found as follows[14]: What is defined by (1) or (2) may be said to be a measure of

---

[12] Notice, furthermore, that by replacing s by $s \& {\sim}s$ in (8) we obtain $\mathrm{p}^*(s \vee {\sim}s) = \mathrm{p}^*(t) + \mathrm{p}^*({\sim}t)$ for any t. If cont is to be nonnegative, p*(tautology) will therefore be the maximum of all the values that p* can have, and the assumption that $\mathrm{p}^*(s \vee {\sim}s) = 1$ is thus truly a normalizing stipulation only.

[13] Somewhat surprisingly, this second sense of relative content seems not to have been discussed extensively in the earlier literature on semantic information.

[14] Ideas closely related to the one which underlies the present section have been discussed briefly in a slightly different form by TÖRNEBOHM [1964] pp. 51–52, and employed by ADAMS [1966] pp. 159–168 (under the title "transmitted information").

the information that a statement h carries concerning the subject matter of which it speaks. As was indicated, it is the amount of uncertainty concerning whatever h is about which we get rid of when we learn that h is true. However, very often this is not all the kind of information we are interested in. Frequently we are not interested in the subject matter of h at all. That is, frequently we are not interested in the reduction in our uncertainty concerning what h says, but rather in the reduction of uncertainty concerning certain other matters which our knowledge of h brings out. These matters may be described by some other statement, e.g. by g. Then the amount of information that h conveys (contains) concerning the subject matter of g is naturally measured by

$$\mathrm{inf}(g) - \mathrm{inf}(g|h) = \log[\mathrm{p}(g|h)/\mathrm{p}(g)] = \log[\mathrm{p}(g \,\&\, h)/\mathrm{p}(g)\mathrm{p}(h)] \tag{11}$$

or, alternatively, by

$$\mathrm{cont}(g) - \mathrm{cont}_{\mathrm{add}}(g|h) = 1 - \mathrm{p}(g \vee h). \tag{12}$$

This may be compared with the analogous expression

$$\mathrm{cont}(g) - \mathrm{cont}_{\mathrm{cond}}(g|h) = \mathrm{p}(g|h) - \mathrm{p}(g). \tag{13}$$

The explanation of (11) is very simple: $\mathrm{inf}(g)$ is precisely the uncertainty which we get rid of when we come to know g; $\mathrm{inf}(g|h)$ is the information g adds to that of h, i.e. the uncertainty that there remains concerning g even after we have learned that h is true. Hence $\mathrm{inf}(g) - \mathrm{inf}(g|h)$ measures the reduction of our uncertainty concerning g which takes place when we come to know, not g, but h. Similar remarks apply to (12).

In the case of (13), somewhat different explanation is called for. Here $\mathrm{cont}(g)$ is the informative content of g a priori, whereas $\mathrm{cont}_{\mathrm{cond}}(g|h)$ is what the informative content of g would be in a world in which we could restrict ourselves entirely to possibilities compatible with h, i.e. in a world in which we know the truth of h. The difference (13) is thus primarily the *change in the information-carrying status of g which takes place when one comes to know h.*

The value of (13) can be either positive or negative. This is in keeping with the explanation just given of the meaning of (13), for coming to know h can obviously affect the relative information-carrying status of g positively or negatively.

It is seen that if we put $h = g$, we obtain from (11) $\mathrm{inf}(g)$ and from (12) as well as (13) $\mathrm{cont}(g)$. This result is of course obvious in view of the intuitive meanings of (11)–(13), respectively. It shows that these new senses of information are generalizations of (1) and (2), respectively.

It is also seen at once that (11) can be either positive or negative, whereas

(12) cannot obtain negative values. This is natural in view of the intuitive difference between the two measures inf and cont. Our *surprise* at finding that g is the case can be greater after we have first found out that h is true than it would be a priori. Thus h may have a negative surprise value concerning the subject matter of g.

In contrast, the substantive information that h conveys concerning the subject matter of any g has to be positive (or zero), as it was found to be.

When g and h are independent (with respect to the probability measure p), (11) is zero. When $h \supset g$ is logically true, $(g \;\&\; h)$ is logically equivalent with h, and (11) therefore receives the value $-\log p(g) = \inf(g)$. As might be expected, in this case the information h carries concerning the subject matter of g is the same g itself carries. This is in fact an optimal case, for if g is constant but h varies, $\inf(g)$ is the largest value (11) can receive.

Similarly, (12) and (13) also assume their maxima (for a constant g) when g is logically implied by h. Furthermore, (13) is like (11) in that it assumes the value $=0$ when g and h are probabilistically independent.

It is also worth noting that both (11) and (12) are symmetric with respect to g and h, whereas (13) is not.

Because of the importance of the ideas that underlie (11)–(13), a special notation may be in order. I shall use for (11) the expression $\text{transinf}(h|g)$, for (12) the expression $\text{transcont}_{\text{add}}(h|g)$, and for (13) the expression $\text{transcont}_{\text{cond}}(h|g)$. The suggested approximate readings are: "the information h conveys concerning the subject matter of g" for the first two and "the change in the content of g due to h" for the third. The motivation for the notation will be given later.

Examples of the distinction between the plain information $\inf(h)$ or $\text{cont}(h)$ and one of the relative senses of information (11)–(13) abound in ordinary life. If I am told that such-and-such weather conditions obtain in San Francisco today (e.g. "it is foggy today"), what is the information this statement (call it "s") gives me? Usually, I am not impressed by the reduction of my uncertainty concerning San Francisco weather today that this statement brings about, i.e., I am not particularly interested in $\inf(s)$ or in $\text{cont}(s)$. I am much likelier to be interested in the information this statement gives me concerning the weather conditions that might be expected to obtain on some other day, e.g. tomorrow. For instance, I might be interested in the information s conveys concerning the chances that it will be foggy in San Francisco tomorrow, too. If this is expressed by the statement t, this means that I am interested in $\text{transinf}(s|t)$ or perhaps in $\text{transcont}_{\text{add}}(s|t)$ or $\text{transcont}_{\text{cond}}(s|t)$ rather than in $\inf(s)$ or in $\text{cont}(s)$.

This homely example perhaps also illustrates the ubiquitousness of the senses of information we have just characterized. Another indication of their prevalence is the role of expressions like (11) in statistical information theory, where they are said to define *transmitted information.*[15] The connection between this usage and the explanations we have given of (11) is straightforward: in the case of the transformation of messages we are interested in the amount of information carried by the statement that a certain definite message is received concerning whether what was sent or not. If h is the statement about the arrival of the message and g the statement about its being sent, then this is precisely what (11) expresses.

6. Expected information

What we have said so far does not exhaust all the senses (or kinds) of information one might legitimately be interested in. Often, in the kind of situation with which my example dealt, we are not primarily interested in the reduction of our uncertainty as to whether it will be foggy in San Francisco tomorrow. Rather, we might be interested in the reduction of our uncertainty as to *whether or not* it will be foggy. One important kind of information can be measured by measuring the reduction of this uncertainty. A way of doing so is as follows: Let $e=$ "it is foggy today" and $h=$ "it will be foggy tomorrow". Then the expected value of the information which e yields concerning the state of affairs described by h or by $\sim h$ is

$$\mathrm{p}(h|e)\,\mathrm{transinf}(e|h) + \mathrm{p}(\sim h|e)\,\mathrm{transinf}(e|\sim h) \tag{14}$$

or

$$\mathrm{p}(h|e)\,\mathrm{transcont}_{\mathrm{add}}(e|h) + \mathrm{p}(\sim h|e)\,\mathrm{transcont}_{\mathrm{add}}(e|\sim h) \tag{15}$$

depending whether we are considering unexpectedness or substantial information.

The expression analogous to (15)

$$\mathrm{p}(h|e)\,\mathrm{transcont}_{\mathrm{cond}}(e|h) + \mathrm{p}(\sim h|e)\,\mathrm{transcont}_{\mathrm{cond}}(e|\sim h) \tag{16}$$

measures the expected change in the information of the true answer to the question: "Is h or $\sim h$ the case?" that is brought about by our knowledge of e.

More generally, if we are dealing with a number of pairwise exclusive and collectively exhaustive alternatives $h_1, h_2, \ldots, h_k$, then the information which

[15] Cf. e.g. SHANNON and WEAVER [1949] pp. 36, 110–111. The difference $\mathrm{inf}(g) - \mathrm{transinf}(h|g) = \mathrm{inf}(g|h)$ is the so-called "equivocation" of the statistical information theorists.

e carries concerning the subject matter they speak of is in the present sense of the word measured by

$$\sum_i \mathrm{p}(h_i|e)\,\mathrm{transinf}(e|h_i) = \sum_i \mathrm{p}(h_i|e)\log\frac{\mathrm{p}(h_i|e)}{\mathrm{p}(h_i)}$$
$$= \sum_i \mathrm{p}(h_i|e)\log[\mathrm{p}(h_i \,\&\, e)/\mathrm{p}(h_i)\,\mathrm{p}(e)] \qquad (17)$$

or by

$$\sum_i \mathrm{p}(h_i|e)\,\mathrm{transcont}_{\mathrm{add}}(e|h_i) = 1 - \sum_i \mathrm{p}(h_i|e)\,\mathrm{p}(h_i \vee e), \qquad (18)$$

respectively. An analogous expression gives the expected effect of our "evidence" e (as we may think of it) on the true alternative h_i:

$$\sum_i \mathrm{p}(h_i|e)\,\mathrm{transcont}_{\mathrm{cond}}(e|h_i). \qquad (19)$$

Expressions of the form (17) occur in statistical information theory and also in theoretical statistics.[16] Although they have not been considered very often by logicians and philosophical methodologists, it seems to me that they ought to have been. They arise on the basis of the very general considerations which were sketched above and which are largely independent of the particular kind of probability which has been assumed to underlie our discussion. They are not due to the statistical character of the ordinary "statistical" information theory nor to any particular sense of probability employed therein.

7. Applications of the distinctions. Local versus global theorizing

The concept of relative information defined by (17) and (18) can be illustrated by means of many different examples. One of them pertains to a line of argument which I used in an earlier paper. Suppose we have a number (say k) of primitive predicates by means of which we can classify individuals into $K=2^k$ Q-predicates (in Carnap's sense). Assume that we are interested in the question how the individuals of our universe of discourse are distributed among the different Q-predicates. More specifically, we may ask whether they leave some Q-predicates empty, i.e. whether any true general laws can be formulated in terms of our primitive predicates and quantifiers, and also what the relative frequencies of the different Q-predicates are. In both these respects, the information conveyed by an observation-statement concerning a finite number of individuals (stating which Q-predicates belong

[16] See e.g. SAVAGE [1954] pp. 50, 235–238.

to them) depends on what we know (or, if we are subjectivists, what we believe) of the overall regularity of our universe.

An extreme case is one in which the universe is known to be completely regular: all individuals have the same Q-predicate, although we do not know a priori which. Then an observation-statement concerning a single individual gives us all the information we are interested in, both information concerning laws and information concerning relative frequencies.

More generally, an observation statement is clearly the more informative the more regular the universe is known to be. For instance, if we know that our individuals are likely to be concentrated heavily on a few Q-predicates, the observation of (say) just two individuals having the same Q-predicate tells us more about the universe than it would do if we knew that the universe is "disorderly" in the sense that its members are likely to be divided evenly among all the Q-predicates. In the latter case, the identity of the Q-predicates which the two individuals have is likely to be dismissed as a freak.

Elsewhere I have briefly discussed the consequences of observations of this kind.[17] Here I want to emphasize only the fact that the sense in which we are here speaking of the information of an observation-statement has to be defined by (17) or (18) rather than by (1) or (2). What is involved is the information these singular observation-statements convey concerning possible generalizations (strict or statistical generalizations). In fact, it is readily seen that the "absolute" information (defined by (1) or (2)) of the observation-statements is not what is involved, for it is low precisely when an observation-statement is highly informative about generalizations, in the sense just explained, i.e. when a high degree of regularity may be expected. This illustrates quite strikingly the difference between (1)–(2) on one hand and (17)–(18) on the other.

It also illustrates the fact that the difference between $\mathrm{inf}(e)$ and $\mathrm{transinf}(e|h)$ (as well as between $\mathrm{cont}(e)$ and $\mathrm{transcont}(e|h)$) is especially important in the case of a singular statement e. We are seldom interested in the information such a statement conveys concerning its own subject matter, but rather in what it tells us concerning other (unobserved) individuals or concerning different possible generalizations. In contrast, we are much likelier to be interested in the information generalizations carry concerning their own subject matter.

However, there are other important differences between different kinds of situations. One of the most important uses that our distinctions have is

[17] See HINTIKKA [1968].

to show that there are several different ways of looking at the relation of observational data to those hypotheses which are based on them and which perhaps are designed to explain them. In different situations the concept of information can be brought to bear on this relation in entirely different ways. There are hence no unique explications of such concepts as "explanation" and "degree of factual (evidential) support". No wonder, therefore, that a host of different explications have been offered for the latter notion in current literature.[18] In general, the scientific search for truth is much less of a single-goal enterprise than philosophers usually realize, and suitable distinctions between different senses of information perhaps serve to bring out some of the relevant differences between different goals.

Let us consider some differences between different cases. One of the most important distinctions here is between, on one hand, a case in which we are predominantly interested in a particular body of observations e which we want to explain by means of a suitable hypothesis h, and on the other hand a case in which we have no particular interest in our evidence e but rather want to use it as a stepping-stone to some general theory h which is designed to apply to other matters, too, besides e. We might label these two situations as cases of local and of global theorizing, respectively. Often the difference in question can also be characterized as a difference between explanation and generalization, respectively. Perhaps we can even partly characterize the difference between the activities of (local) explanation and (global) theorizing by spelling out (as we shall proceed to do) the difference between the two types of cases.

8. Maximum likelihood principle and its presuppositions

In the former case, we want to choose the explanatory hypothesis h such that it is maximally informative concerning the subject matter with which e deals. Since we know the truth of e already, we are not interested in the substantive information that h carries concerning the truth of e. What we want to do is to find h such that the truth of e is not unexpected, given h. This means that we want to deal with the measure of unexpectedness inf rather than with the measure cont, and to choose h such as to reduce the surprise value of e as much as possible.

Thus we arrive at the suggestion that in the case of explanation (local

[18] A convenient summary of a number of them is provided in the survey article by KYBURG [1964]. Cf. also GOOD [1960] and TÖRNEBOHM [1966].

theorizing) we should choose h so as to maximize

$$\mathrm{transinf}(h|e) = \log[\mathrm{p}(e|h)/\mathrm{p}(e)]. \qquad (20)$$

Since e was assumed to be constant, to maximize (20) means to choose h so as to maximize the conditional probability $\mathrm{p}(e|h)$, known as the likelihood of e given h. Thus we arrive at the famous maximum likelihood principle as the natural method of choosing one's explanatory hypothesis in the kind of circumstances indicated (local theorizing).[19] Thus the importance of this principle in statistics has an interesting general reason which can be brought out in terms of the concept of information.

Törnebohm has suggested using (20), suitably normalized, as a measure of the evidential strength of h vis-à-vis e.[20] In view of what we just found, it is not surprising that he should end up with the maximal likelihood principle when he applies the principle to a simple case.

At the same time our observations may indicate some of the limitations of Fisher's maximum likelihood principle. Its rationale was seen to be in terms of the concept of information. This recalls Fisher's insistence that he is not dealing with inductive behavior or with decisions under uncertainty, but with "the improvement of natural knowledge".[21] Our description above of the circumstances in which we are led to the maximum likelihood principle also recalls the kind of situation (analysis of experimental data) which Fisher typically deals with.[22] It may be that the maximum likelihood method is tied more closely than its proponents sometimes realize to cases where the explanation of the particular data predominates over other concerns. It appears that the maximum likelihood principle is a weapon of explanation rather than of generalization.

This does not mean that the maximum likelihood principle cannot occasionally be quite useful for the latter purpose, too. This is the case when we already know that any explanation of the particular data we have will serve to explain whatever other data we might be interested in. This presupposes, obviously, that the regularity of our universe is already known to be maximal (or very great) in the relevant respects – in the sense that whatever observations we make concerning a part of it can be carried over intact so as to

---

19 For the principle, see e.g. Cramér [1946] pp. 498–506; Fisher [1956] pp. 68–72.

20 Törnebohm [1966].

21 Fisher [1956] pp. 100–104.

22 In Fisher's view, "the science of statistics is essentially a branch of Applied Mathematics, and may be defined as mathematics as applied to observational data" (Fisher [1925] p. 1).

apply to others.[23] However, to paraphrase a well-known statement of Mill's, whoever answers the question as to when this is possible, has solved the riddle of induction, and the maximum likelihood principle does not help us in that enterprise. These observations are connected with Fisher's penchant to assume, in the case of estimation, that the correct form of a law has already been discovered and that there remains only the task of estimating the numerical values of the parameters these laws involve. What I would consider the main aspect of theorizing as distinguished from explanation, viz. determining the form of these laws, is thus assumed to have been accomplished before the estimation problem is raised.

In another paper, I have studied some ways of expressing the regularity which we have been discussing by means of explicit parameters.[24] Whenever the regularity is not maximal, we have to be wary of generalizing too much from observations. The strength of this conservatism varies inversely to the regularity we are allowed to assume there is in the universe. Carnap's λ is a well-known index of conservatism of this kind. I have argued that its uses are restricted to singular inductive inference and that we need a comparable index of caution for inductive generalization, too. Be this as it may, it is interesting to see that the maximum likelihood principle results from the types of inductive logic which are calculated to be models (however over-simplified) of genuine inductive inference if and only if $\lambda \to 0$, i.e. only if the regularity of our universe is maximal. It is only then that the maximal likelihood principle is an acceptable tool of inductive inference.

In order to have a firm grasp of the nature of this principle, it is perhaps useful to have a closer look at its consequences for the kind of "Carnapian" situation which has already been used earlier as a testing ground (applied monadic first-order language). In such a situation, we can classify observed individuals into a number of classes (some of which may be empty) and which form a partition of our universe of discourse. (They may be given by the Q-predicates mentioned above.) If we have observed a random sample of individuals belonging to some of these, what hypothesis should we set up

[23] Is it accidental that so many of the most successful applications of Fisherian techniques have been in the field of biology? Is it perhaps the case that the presuppositions of the use of these techniques for the purpose of genuine theorizing are likelier to be satisfied there than in the social sciences? Once you have found out all the relevant features of a few well-developed members of a species, you are apt to have found out what you wanted for the whole species, one is tempted to say, and this would be precisely what the assumption of maximum regularity amounts to.

[24] See HINTIKKA [1966] and cf. Carnap's earlier work on the index of caution λ for singular inductive inference, especially CARNAP [1952].

concerning the whole universe? The answer which the maximum likelihood principle gives is known as the straight rule.[25] It tells us to assume that the actual relative frequencies of the different kinds of individuals in the whole universe are the same as the observed relative frequencies.

It is not hard to see that this is not a very realistic method of choosing our hypothesis in many cases, especially when our sample is small. (Suppose e.g. that we have observed only two individuals of different kinds. It does not appear motivated to assume on the basis of such a sample that all the individuals of the universe belong to the two kinds our pair of observed individuals exemplify, or that they are divided precisely evenly between the two.) Hence the straight rule is rejected by the majority of writers on induction as a method of generalizing from random samples. It is important to realize, however, that this rejection depends entirely on our interest in what they hypothesis we adopt says of individuals other than the observed ones, in brief, what its total information is. If our sole purpose were merely to give the best possible available explanation of our observations, we might be led to change our preferences and opt for the straight rule. And we would certainly do so in any case if we happened to know that the universe is completely regular.

An interesting interpretation of (11) is obtained by observing that

$$\text{transinf}(h|g) = \inf(g) - \inf(g|h) = \inf(g) + \inf(h) - \inf(g \,\&\, h)$$

which can obviously be interpreted as the amount of information shared by h and g. Starting from this idea, Törnebohm makes the shrewd observation that the use of (20) (perhaps suitably normalized) as a measure of the acceptability of h seems to be in good agreement with the practice of historians.[26] A historian, clearly, is striving to organize his narrative so as to make the information it conveys overlap as closely as possible with the information his sources (i.e. the evidence e) contain, which is just what the maximization of (20) (for a fixed e) amounts to. It is important to realize, however, that in this respect the interests of a historian are apt to differ from those of a scientist. In the recent discussions concerning the methodology of history, it has frequently been insisted that a historian does not strive to generalize.[27] Part of what is perhaps meant by such statements can now be appreciated. A historian is in the first place trying to explain the particular data he has,

---

[25] Cf. CARNAP [1952].

[26] TÖRNEBOHM [1966] p. 85.

[27] See e.g. DRAY [1956] and the references to the subsequent discussion in PASSMORE [1966] pp. 539–540.

and is not overtly concerned with setting up theories with a view to saying something informative about individuals and events not immediately relevant to these data.[28] This is precisely why a historian can be said (very roughly of course) to be guided, in his choice of a hypothesis h, by the expression (20) for the information which h gives concerning e, rather than by some function that turns on the overall information of h.

More generally, the importance of likelihoods in statistical testing can be appreciated in the light of our observations. If two hypotheses h_i and h_j are compared with respect to their informativeness for certain particular data e, we have

$$\text{transinf}(h_i|e) - \text{transinf}(h_j|e) = \log[\text{p}(e|h_i)/\text{p}(e|h_j)]. \qquad (21)$$

In other words, if this is all that matters, the comparison between h_i and h_j turns entirely on the likelihood ratio $\text{p}(e|h_i)/\text{p}(e|h_j)$.

9. Maximizing expected content

However, if we are interested in our data (say e) merely as partial evidence for a global theory h, the situation is quite different. Here we are not so much interested in maximizing transinf$(h|e)$ as maximizing in some sense the information of h itself. Especially when h is a general statement, we are likelier to be interested in the substantive information it yields than in its surprise value. (Trying to maximize substantive information is typically an indication of serious global theorizing.) To maximize this directly means simply to opt for the least likely theory without considering any kind of evidential support, which is obviously unrealistic. The natural course in this case is therefore to consider the expected value of this substantive information as a result of the adoption of the hypothesis (theory) h. If h is true, we gain the (substantive) information cont(h) by adopting h. If h is false, we loose the information which we could have gained by opting for $\sim h$, rather than h. According to what was just said, our net utility in this case is therefore $-\text{cont}(\sim h)$. The expected value of our information gain is thus

$$\text{p}(h|e)\cdot\text{cont}(h) - \text{p}(\sim h|e)\cdot\text{cont}(\sim h), \qquad (22)$$

[28] In view of what was said above of regularity and maximum likelihood, this is not incompatible with saying that general laws are involved in historical explanations, too, provided these general laws are taken to be of a very unproblematic character – so unproblematic, indeed, that if they are known to apply in one case, they can be applied to all the others. This is not entirely unlike what the defenders of "covering-law" explanations in history have been led to say of the generalizations involved in historical explanations.

which simplifies as

$$\mathrm{p}(h|e) - \mathrm{p}(h). \tag{23}$$

This is simply the increase in the probability of h which has been brought about by the evidence e. If expected information is used as a measure of the acceptability of a theory, this acceptability does not depend on a priori or a posteriori probability alone, but on the effect of evidence – on "how well the theory has stood up to the test of experience".[29]

It has been shown elsewhere that a policy of maximizing (23) when one is carrying out a generalization (i.e. choosing h among all the available general statements) results in certain interesting cases in a sound method of inductive generalization.[30]

On the basis of what was said above it is not surprising that it is just in the case of inductive generalization that this strategy succeeds.

Hempel has proposed that we use the relative measure $\mathrm{cont}_{\mathrm{add}}(h|e) = \mathrm{cont}(e \supset h)$ rather than the absolute one $\mathrm{cont}(h)$.[31] The most straightforward way of doing so fails, however, for if we identify the utility of the correct adoption of h with $\mathrm{cont}_{\mathrm{add}}(h|e)$ and the disutility of an incorrect adoption with $\mathrm{cont}_{\mathrm{add}}(\sim h|e)$ (i.e. with the utility which we could have gained by adopting $\sim h$ instead of h), the expected value of our utility will be

$$\mathrm{p}(h|e)\,\mathrm{cont}_{\mathrm{add}}(h|e) - \mathrm{p}(\sim h|e)\,\mathrm{cont}_{\mathrm{add}}(\sim h|e) = \\ = \mathrm{p}(h|e)\,\mathrm{p}(\sim h \,\&\, e) - \mathrm{p}(\sim h|e)\,\mathrm{p}(h \,\&\, e) = 0. \tag{24}$$

This result may have been instrumental in leading Hempel to adopt a different measure of the disutility of an incorrect adoption of h.

An attempt to use $\mathrm{cont}_{\mathrm{cond}}$ instead of $\mathrm{cont}_{\mathrm{add}}$ does not help, for

$$\mathrm{p}(h|e)\,\mathrm{cont}_{\mathrm{cond}}(h|e) - \mathrm{p}(\sim h|e)\,\mathrm{cont}_{\mathrm{cond}}(\sim h|e) = \\ = \mathrm{p}(h|e)\,(1 - \mathrm{p}(h|e)) - (1 - \mathrm{p}(h|e))\,\mathrm{p}(h|e) = 0.$$

The use of (23) as a function to be maximized derives further support from other considerations. It lies close at hand to suggest that we should maximize, not the expected absolute content $\mathrm{cont}(h)$ of a hypothesis h, but rather the

[29] The similarity between this characteristic of (23) and what Popper says of his notion of corroboration might prompt one to offer (23) as a possible explication of corroboration. Good has in effect done this, in the highly interesting paper GOOD [1960]. Actually, there is also a rather close connection between (23) and a slightly different notion of POPPER'S [1954], called explanatory power. (See below, p. 329.)

[30] HINTIKKA and PIETARINEN [1966].

[31] HEMPEL [1962] especially pp. 153–156; HEMPEL [1960].

content it has relative to the subject matter of the conjunction of all evidential statements e, or $\text{transcont}_{\text{add}}(h|e)$. The expected value of this is, according to the principles already used in obtaining (22),

$$\text{p}(h|e)\,\text{transcont}_{\text{add}}(h|e) - \text{p}(\sim h|e)\,\text{transcont}_{\text{add}}(\sim h|e). \tag{25}$$

Substituting for $\text{transcont}_{\text{add}}(h|e)$, which by symmetry is $\text{transcont}_{\text{add}}(e|h)$, the expression $\text{cont}(h) - \text{cont}_{\text{add}}(h|e)$ we obtain from (25)

$$[\text{p}(h|e)\,\text{cont}(h) - \text{p}(\sim h|e)\,\text{cont}(\sim h)] + \\ - [\text{p}(h|e)\,\text{cont}_{\text{add}}(h|e) - \text{p}(\sim h|e)\,\text{cont}_{\text{add}}(\sim h|e)].$$

Here the second term is (24), and hence zero, while the first is (22) and hence identical with (23) or $\text{p}(h|e) - \text{p}(h)$. Thus we obtain the same result no matter whether we are trying to maximize the expected value of the absolute content $\text{cont}(h)$ of h or the expected value of the content $\text{transcont}_{\text{add}}(h|e)$ of h with respect to the evidence e. This result may seem surprising. It is understood, however, when we realize that the only "boundary condition" which prevents us from maximizing the content of h ad libidum is the requirement that it has to agree with the evidence e we have. Thus the task in both cases is essentially to maximize the content (substantive information) that h gives us concerning our evidence.

Furthermore, (23) is according to (13) precisely $\text{transcont}_{\text{cond}}(e|h)$, that is, the effect of evidence e on the information-carrying status of hypothesis h. In choosing h so as to maximize the expected value of its information-content $\text{cont}(h)$, we are ipso facto choosing h so as to maximize the gain which our evidence e gives us concerning the information h carries. There is a certain partial symmetry between the principle involved here and the maximum likelihood principle. In the latter, we are maximizing the information (in the sense of (11)) which the hypothesis gives us concerning the evidence, whereas in the former we are maximizing the gain in information (change in the information-carrying status) which the given evidence e brings about concerning a hypothesis.

But why maximize $\text{transcont}_{\text{cond}}(e|h)$ and not its expected value? This contrast is empty, for according to the principles used in arriving at (22) the expectation of $\text{transcont}_{\text{cond}}(e|h)$ is

$$\text{p}(h|e)\,\text{transcont}_{\text{cond}}(e|h) - \text{p}(\sim h|e)\,\text{transcont}_{\text{cond}}(e|\sim h) = \\ = \text{p}(h|e)\,[\text{p}(h|e) - \text{p}(h)] - [1 - \text{p}(h|e)]\,[1 - \text{p}(h|e) - (1 - \text{p}(h))] = \\ = \text{p}(h|e) - \text{p}(h) = \text{transcont}_{\text{cond}}(e|h). \tag{26}$$

In other words, to maximize $\text{transcont}_{\text{cond}}(e|h)$ is to maximize its expected value, giving us for the fourth time the same expression to be maximized.

10. Degrees of "evidential support"

Thus (23) serves as an especially interesting index of the relation of a hypothesis h to the evidence e. In suitable circumstances, it might be considered as an index of the acceptability of h on the basis of e. Measures of such acceptability are sometimes known as indices of "evidential power" of e with respect to h or the "corroboration" that e lends to h.[32] If (23) is used as a measure of evidential power, it may be desirable to normalize it in a suitable way so as to make the bounds of its variation more uniform.

When h and e are independent with respect to the probability measure p, (23) is zero, for then $\text{p}(h|e)=\text{p}(h)$. When e logically implies h, $\text{p}(h|e)=1$, and (23) reduces therefore to $\text{p}(\sim h)$. When e logically implies $\sim h$, (23) reduces to $-\text{p}(h)$. It lies here close at hand to multiply (23) by a suitable normalizing factor. Among the available possibilities there are e.g. the following normalizing factors: (a) $1/\text{p}(h)\,\text{p}(\sim h)$; (b) $1/\text{p}(h)$; (c) $\text{p}(e)/\text{p}(\sim h)$. These give us the following respective values in the two extreme cases mentioned above: (a) $1/\text{p}(h)$ and $-1/\text{p}(\sim h)$; (b) $\text{p}(\sim h)/\text{p}(h)$ and -1; (c) $\text{p}(e)$ and $-\text{p}(e)\,\text{p}(h)/\text{p}(\sim h)$.

These still depend on h, however. In order to see one way of making them independent of h, let us first use the normalizing factor (a) which is perhaps the most natural one here. The result can easily be rewritten as

$$\frac{\text{p}(e|h)-\text{p}(e|\sim h)}{\text{p}(e)} \tag{27}$$

which now varies between $1/\text{p}(h)$ and $-1/\text{p}(\sim h)$. An obvious way to "normalize" this so as to vary between $+1$ and -1 is to replace it by

$$\frac{\text{p}(e|h)-\text{p}(e|\sim h)}{\text{p}(e|h)+\text{p}(e|\sim h)}. \tag{27*}$$

This always changes in the same direction as (27).

It behaves in an orderly way: it varies between $+1$ and -1 and receives these two extreme values only when $\vdash(e\supset h)$ and $\vdash(e\supset \sim h)$, respectively, as one can readily verify.

Most of the expressions which we have recently mentioned have actually

[32] For a possible difference between these two notions, see GOOD [1960].

been proposed by different writers as measures of evidential power. A number of such suggestions have been conveniently summarized by Kyburg.[33] Of the proposals he lists, those by Levi and Carnap amount to (23) itself. (No suggestion to this effect seems to have been actually made by Carnap, however, and Levi therefore appears to have been the first philosopher to use (23) as a measure of acceptability of h relative to e.) The use of the normalizing factor (b) is tantamount to a suggestion by Finch, and the use of (c) to one by Rescher, while (27)* has been proposed by Kemeny and Oppenheim as a measure of the degree of factual support. If we had normalized (23) directly, without first rewriting it as (27), that is to say, if we had replaced it simply by

$$\frac{\mathrm{p}(h|e) - \mathrm{p}(h)}{\mathrm{p}(h|e) + \mathrm{p}(h)} \tag{28}$$

we would have obtained Popper's measure of the explanatory power of e with respect to h. (The limits of its variation are $(1-\mathrm{p}(h))/(1+\mathrm{p}(h))$ and -1, the former of which still depends on h.) From the point of view here adopted, all these proposed definitions thus turn out to be closely related to each other, and also turn out to be partly justifiable in terms of the concept of information. There is thus more uniformity in different philosophers' conceptions of factual support (or explanatory power) than first meets the eye. It is perhaps significant that all the proposals we have mentioned have turned out to be closely connected with the concept of substantive information (the expected value of our cont) rather than with the concept of surprise value (unexpectedness).

11. Possible reasons for preferring one sense of information

I said earlier that scientific inquiry is a multi-goal enterprise and that there cannot therefore be any uniform measure of its success, i.e. any unique measure of how "good" one's hypotheses and explanations are. This does not mean, however, that the choice of our measures of success – which normally prejudges partly our choice of the statistical techniques involved – is arbitrary or a matter of convention. Very often there are some objective guide-lines, and the realization of what one's objectives are and what their

[33] See KYBURG [1964], who supplies further references to the relevant literature. It is perhaps worthwhile to note that in Kemeny and Oppenheim an important role is played by the requirement that their measure of factual support must vary between -1 and $+1$. See KEMENY and OPPENHEIM [1952].

realization presupposes can be a matter of considerable urgency. For instance, if a social science aims at a genuine theory, it cannot be satisfied with explanations of the different particular data that it has available, however "rich" these data and these explanations may be. For a discussion of this point in the case of sociology, see Bernard P. Cohen, "On the construction of sociological explanations" (forthcoming).

There is another, much vaguer criterion that might also be worth a closer scrutiny. Often, the best way of describing an inductive situation is to think of it as arising from a second-order probability distribution, that is, from a probability distribution on a class of probability distributions. De Finetti's famous representation (basis) theorem in fact says that this can be done under quite weak conditions. Now it may be the case that we do not know a priori some of the characteristic parameters of this second-order distribution, but have to estimate them on the basis of our evidence. How are we to do this? In some vague sense, it seems that for this purpose the maximum likelihood principle is appropriate even where it is clearly inappropriate for the purpose of estimating the characteristics of the (true) first-order distribution. In so far as this feeling can be rationalized, it seems to turn on the fact that the second-order distribution exhaust our interest in the world. It may often be inappropriate to focus our explanation on one particular set of data, for there may be other sets of data available which we also have to heed in developing a satisfactory overall theory. But in the case of an explanation of the global features of the universe (as it were), there is nothing else we could possibly be interested in, and hence we can be happy if we can explain them to the exclusion of everything else, for there are no other universa which we have to beware of. And this of course means that the maximum likelihood principle is appropriate.[34]

References

Adams, E.W., On the nature and purpose of measurement, Synthese **16** (1966) 125–169.
Bar-Hillel, Y., Language and information, selected essays on their theory and application (Addison-Wesley Publ. Co., Inc., Reading, Mass., 1964).
Bar-Hillel, Y. and R. Carnap, Semantic information, Brit. J. Phil. Sci. **4** (1953) 144–157.
Carnap, R., Continuum of inductive methods (Chicago, University of Chicago Press, 1952).
Carnap, R. and Y. Bar-Hillel, An outline of a theory of semantic information, Techn. Report no. 247 (Research Laboratory of Electronics, Massachusetts Institute of Technology, 1952); reprinted in: Bar-Hillel [1964].

[34] It would be interesting to examine whether this is perhaps part of the reason why Reichenbach resorted to higher-order probabilities in his defense of the straight rule; see Reichenbach [1949].

CHERRY, C., On human communication (The M.I.T. Press, Cambridge, Mass., 1957; second edition, 1966).

COX, R.T., The algebra of probable inference (The John Hopkins Press, Baltimore, 1961).

CRAMÉR, H., Mathematical methods of statistics (Princeton, Princeton University Press, 1946).

DRAY, W., Laws and explanation in history (Oxford, Clarendon Press, 1956).

DUBINS, L.E. and L.J. SAVAGE, How to gamble if you must: inequalities for stochastic processes (New York, McGraw-Hill, 1965).

FISHER, Sir R.A., Statistical methods for research workers (Edinburgh, Oliver and Boyd, 1925).

FISHER, Sir R.A., Statistical methods and scientific inference (Edinburgh, Oliver and Boyd, 1956).

GOOD, I.J., Weight of evidence, corroboration, explanatory power, information and the utility of experiments, J. Roy. Stat. Soc. B **22** (1960) 319–331.

HANNA, J., A new approach to the formulating and testing of learning models, Synthese **16** (1966) 344–380.

HEMPEL, C.G., Deductive-nomological versus statistical explanation, in: Scientific explanation, space and time, Minnesota studies in the philosophy of science, Vol. 3, eds. H. Feigl and G. Maxwell (Minneapolis, University of Minnesota Press, 1962) pp. 98–169.

HEMPEL, C.G., Inductive inconsistencies, Synthese **12** (1960) 439–469; reprinted in: C.G. Hempel, Aspects of scientific explanation and other essays in the philosophy of science (New York, The Free Press, 1965) pp. 53–79.

HINTIKKA, J., Towards a theory of inductive generalization, in: Logic, methodology and philosophy of science, ed. Y. Bar-Hillel (Amsterdam, North-Holland Publ. Co., 1965) pp. 274–288.

HINTIKKA, J., A two-dimensional continuum of inductive logic, in: Aspects of inductive logic, eds. J. Hintikka and P. Suppes (Amsterdam, North-Holland Publ. Co., 1966) pp. 113–132.

HINTIKKA, J., On semantic information, to appear in: Proc. Intern. Colloq. on logic, physical reality and history at the University of Denver, ed. W. Yourgrau (The Plenum Press, New York, 1968).

HINTIKKA, J. and J. PIETARINEN, Semantic information and inductive logic, in: Aspects of inductive logic, eds. J. Hintikka and P. Suppes (Amsterdam, North-Holland Publ. Co., 1966) pp. 96–112.

KEMENY, J.G., A logical measure function, J. Symb. Logic **18** (1953) 289–308.

KEMENY, J.G. and P. OPPENHEIM, Degree of factual support, Phil. of Sci. **19** (1952) 307–324.

KHINCHIN, A.I., Mathematical foundations of information theory (Dover Publications, N.Y., 1957).

KYBURG, H.E., Recent work in inductive logic, Am. Phil. Quart. **1** (1964) 249–287.

PASSMORE, J., A hundred years of philosophy (London, second edition, 1966).

POPPER, K.R., Logik der Forschung (Springer-Verlag, Wien, 1935); transl. with new notes and appendices as: The logic of scientific discovery (Hutchinson and Co., London, 1959).

POPPER, K.R., Degree of confirmation, Brit. J. Phil. Sci. **5** (1954) 143–149. (Correction ibid. 334.)

REICHENBACH, H., The theory of probability (Berkeley and Los Angeles, University of California Press, 1949).

SAVAGE, L.J., The foundations of statistics (New York, John Wiley and Sons, 1954).

SHANNON, C.E. and W. WEAVER, The mathematical theory of communication (The University of Illinois Press, Urbana, Illinois, 1949).

TÖRNEBOHM, H., Information and confirmation, Gothenburg studies in philosophy, Vol. 3 (Stockholm, Almquist and Wiksell, 1964).

TÖRNEBOHM, H., Two measures of evidential strength, in: Aspects of inductive logic, eds. J. Hintikka and P. Suppes (Amsterdam, North-Holland Publ. Co., 1966) pp. 81–95.

11

ON THE INCOMMENSURABILITY OF THEORIES*

The commensurability of two theories can be defined (relative to a given set of questions) as the ratio of the total information of their shared answers to the total information of the answers yielded by the two theories combined. Answers should be understood here as model consequences (in the sense of the author's earlier papers), not deductive consequences. This definition is relative to a given model of the joint language of the theories, but can be generalized to sets of models. It turns out to capture also the idea of incommensurability as conceptual alienation. Incommensurability so defined does not imply incomparability.

1. Incommensurability as Unstranslatability. One of the most intriguing ideas in recent philosophy of science is the notion of incommensurability as applied to scientific theories.[1] It is, for instance, one of the most central tenets of Thomas Kuhn's approach to philosophy of science that genuinely scientific theories dealing with the same phenomena can be incommensurable. In spite of its central role, this notion has not been subjected to a satisfactory conceptual analysis. As a consequence, philosophical arguments involving the notion of incommensurability are hard to evaluate and sometimes even fallacious. An example of this fallaciousness has been pointed out elsewhere. It has been shown that what probably is the one most frequently used (and apparently the oldest) specific type of example of the alleged incommensurability of scientific theories depends on mistaken ideas of the nature of the identification of concepts by means of the theories in which they play a role.[2]

*Received May 1986; revised August 1986.

†This work has made possible by NSF Grant #IST-8310936 (Information Science and Technology).

[1]This paper is in effect a part of a larger work tentatively entitled *An Interrogative Model of Inquiry and Some of Its Applications*. For the approach used there, see Hintikka (1984a, 1984b, 1985), and Hintikka and Hintikka (1982).

[2]It is pointed out in the work mentioned in note 1 above that a theory may perfectly well allow one of its concepts to be identified or measured in a different way in its different models (applications). Hence the widespread type of argument that purports to establish the incommensurability of two theories by pointing out that in them some key concept is identified or measured in different ways cuts no ice whatsoever, for the same can be true of different applications of the very same theory.

Some philosophers apparently mean by the incommensurability of two theories, say T_1 and T_2, little more than the impossibility of comparing the merits of T_1 and T_2 as scientific theories in a purely objective and value-free way. In that sense, a discussion of incommensurability will amount to a discussion of different methods of theory comparison, of their presuppositions, and especially of their role in actual scientific practice. But if this is all that is meant by incommensurability, the concept does not play any functional role in arguments intended to establish, for example, the impossibility of a rational evaluation of scientific theories. For to argue that theories are incomparable because they are incommensurable would then amount to arguing that they are incomparable because they are incomparable.

For this reason, philosophers employing the concept of incommensurability are well advised to try to provide a further analysis of it. Even though full analyses are conspicuously absent from recent philosophical discussion, a somewhat more specific sense is nevertheless often associated with the notion of incommensurability. It is taken to mean some kind of conceptual relation, or perhaps lack of a relation, between the two theories in question, a relation that is supposed to explain why they are incomparable. In this paper, I shall locate one particular relationship that seems to me to capture at least some of the ideas that have been intended by those philosophers and historians of science who have employed the notion of incommensurability. Since I am not using the term as a mere synonym of incomparability, my explication will initially leave open the question whether or not incommensurable theories in the sense defined can be rationally compared with each other for their scientific merits.

It might *prima facie* seem surprising that there is no single sharp logical explication of the notion of incommensurability in the literature. The way in which the notion is usually characterized nevertheless explains why philosophers have thought that no precise explication of incommensurability using the tools of logic and formal semantics is possible. For incommensurability is typically thought of as a radical conceptual discrepancy between two theories, akin to Quine's infamous idea of the impossibility (underdeterminacy) of radical translation. Indeed, the incommensurability of T_1 and T_2 has been construed as something like an impossibility of translating one of those theories into the language of the other and of translating both of them into some third "theory-neutral" language (Laudan 1977, p. 42). If so, an explicit logico-linguistic definition of incommensurability appears out of the question; for how could such a definition relate to each other two theories that by assumption are conceptually unrelatable?

Both this rhetorical question, and the line of thought that gave rise to it, appear to me fundamentally mistaken. Even if we think of incommensurability as conceptual alienation, there is no reason to think that such conceptual discrepancy of T_1 and T_2 could not manifest itself in other ways, even in ways that would enable us to define the notion in an interesting way. As a consequence, there is nothing that rules out a logical explication of the concept, either. I shall argue that the interrogative model of inquiry which I have outlined elsewhere provides a concrete and precise definition of at least one interesting kind of incommensurability (Hintikka 1984a, 1984b, 1985; Hintikka and Hintikka 1982).

2. Pooling Together Two Languages. First, we can note that the requirement of intertranslatability is in any case not a necessary condition of commensurability; in other words, that it is far too strong a requirement in some cases. The problem of translation between quantum physics and classical physics is a difficult one, as illustrated by the vast and inconclusive literature on the principle of correspondence. Yet as a matter of brute historical fact, quantum mechanics not only was compared with classical mechanics but was generally accepted in preference to the latter. Hence a translatability can scarcely be a precondition of commensurability.

Moreover, it is simply false that a difference between the respective conceptual frameworks (languages) of two theories T_1 and T_2 always makes them incomparable. There is in principle nothing that prevents a philosopher of science from pooling together the resources of the two conceptual frameworks so as to bring T_1 and T_2 under the same roof conceptually. Admittedly, this is not accomplished by a *fiat* but requires as its prerequisite an analysis of possible interconnections of the two conceptual frameworks.

Ironically, if the two theories are not implicitly related, this analysis can be dispensed with, and a simple union of the conceptual resources of T_1 and T_2 does the trick. What is usually thought of as the hardest case thus turns out to be the easiest.

The real difficulty is that the two conceptual frameworks may be related to each other tacitly in a way which neither theory spells out. Joining the two frameworks together hence often involves the task of teasing out the hidden links between the two frameworks. In some cases, there does not even have to be much that is hidden there. For instance, sometimes one theory, including its basic concepts, is supposed to be a limiting case of the other one in some sense, which may not be obvious. In other cases, the two may simply have some concepts in common. Typically, the common concepts are observation concepts, while the theoretical concepts used by the two theories are largely or totally different.[3]

Thus an adequate pooling together of the two conceptual frameworks may or may not be a straightforward task, but it is not an impossible or senseless task. On the contrary it may be one of the tasks that a practicing scientist or a philosopher of science may be expected to perform.

3. An Explication of Incommensurability. In view of such considerations, let us assume, for the sake of argument, that two theories T_1 and T_2 are formulated in the same conceptual framework, that is, can be codified in a common language. What can it mean then to say that they are incommensurable? Clearly, it means that their respective consequences are somehow so different that they do not enable a scientist to compare them with each other. But what kinds of consequences? It is fairly obvious that purely deductive consequences are not what is intended. It is not clear that any propositions concerning observable particulars are deductively implied by each theory. And when such particular consequences are deducible from a general theory, it is not clear that such consequences automatically tell us anything very much about the theory. So-called paradoxes of confirmation illustrate this.

It seems to me that it is much better to use as a touchstone of incommensurability the set of questions (problems) which a theory can help an inquirer to answer (solve), in the sense of the interrogative model of inquiry defined and studied in my earlier papers (see note 1). This has the interesting and helpful consequence that the incommensurability of two theories, say of T_1 and T_2, will not be an absolute notion, but depends on the model ("world") M in which the interrogative "game" takes place.

But how can the questions which a theory helps to answer be used as a clue to the incommensurability of theories? Let's first see what the situation would be on a strict hypothetico-deductive conception of science. The merits of two competing theories, T_1 and T_2, must on this view be judged, at least partly, on the basis of their observable deductive consequences. If these are not comparable, in other words, if there is no observational proposition S such that T_1 implies S while T_2 implies either S or $\neg S$, there is in the hypothetico-deductive model no basis of adjudicating between the two theories, except perhaps in terms of their overall

[3]The orthodox objection to what I am saying here is that this communality of observational terms is made into a merely apparent one by the concept-ladenness and theory-ladenness of observations. For instance, Kuhn relies heavily on the concept-ladenness and theory-ladenness of observations in his arguments for the incomparability of theories. Arguments of this kind have little force, however, before a satisfactory explication of these mystical ladenness properties has been offered. Elsewhere (forthcoming), I have presented an analysis of these notions that does not imply that concept-ladenness and theory-ladenness of observations should be relevant at all to the issue of theory comparability. Hence I am disregarding this objection here; I believe that it is a red herring.

(potential) strength and simplicity. In the circumstances just envisaged, the observable (deductive) consequences of T_1 and T_2 cannot be compared with one another.

Now it can be argued that the notion of deductive consequence must be replaced by the notion of model consequence.[4] In light of the line of thought just sketched, T_1 and T_2 are (totally) incommensurable if and only if there is no question Q_i in a set of relevant questions $\{Q_i\}$, $i \in q$, which is answerable on the basis of both of the two theories. This, then, is the explication of the notion of the incommensurability of theories which the interrogative model suggests. I propose to call the notion so defined *consequential incommensurability*. Without wanting to deny that other explications are perhaps viable, I believe that this characterization captures most of the pretheoretical ideas of those philosophers, historians of science, and scientists who have spoken of the incommensurability of theories. This characterization was inspired by the hypothetico-deductive conception of science, but it is of course independent of it.

4. Explanations and Consequences. It is relevant to note that consequential incommensurability can in principle obtain between T_1 and T_2 even if they are right from the beginning formulated in the same language. It is not likely that anyone will ever find a historical example of this phenomenon, for different competing theories are normally designed to account for the same problems, and hence to have commensurable consequences. Yet the possibility of such an eventuality is of considerable philosophical interest here, for it shows in effect that the translation criterion mentioned above does not even provide a sufficient condition of commensurability.

On the other hand, it is seen that the idea of pooling together the languages of two possibly incommensurable theories proposed above in section 2 does not prejudge any philosophical points, if my explication of incommensurability is adopted. For assume that the two theories are unrelated in the sense that the conceptual systems codified in their respective languages are disparate. Then pooling them together does not force us to dig up any hidden interrelations (meaning postulates) to link the two languages. Hence the model consequences of the two theories will remain different, being expressed in different and unrelated terms. In such a case, the only consequence of the pooling together of the two languages is that translational incommensurability, far from disappearing, is transformed into consequential incommensurability. This suggests, tentatively, that the

[4]For a definition of the concept of model consequence and for an argument supporting the replacement of the notion of deductive consequence, see the work cited in note 1 above, chapter 2. Essentially, C is a model consequence of T in M iff it can be derived from T in the interrogative game played on M.

latter notion can thus be used to capture the essence of the former too. (Compare with section 6, below.)

As was pointed out, my characterization of incommensurability (consequential incommensurability) is initially relative to some particular model M. It can be generalized by considering the questions that T_1 and T_2 enable the Inquirer to answer in the intended applications (models) M_i of T_1 ($i \in I$) and M_j of T_2 ($j \in J$), respectively. The two are incommensurable if and only if there is no model M_k that is an intended application of both theories ($k \in I$, $k \in J$) and in which T_1 and T_2 enable the Inquirer to answer one and the same relevant question Q_i, $i \in q$.

From this extended definition it follows at once that an important source of the incommensurability of two theories T_1 and T_2 is that their intended applications (intended models) are partly or wholly different. This idea is familiar from recent discussions of incommensurability, but it obviously cannot be captured by any translatability criterion.

The definition of incommensurability just outlined can be formulated more precisely and generalized in various ways. Before doing so, however, a few remarks on the philosophical consequences of what has been found are in order.

For instance, one can see the correctness of Larry Laudan's point that translational incommensurability does not imply that comparisons between the two theories in terms of their question-answering (problem-solving) power are impossible (Laudan 1977, pp. 242–244). If the intended variety of incommensurability is my consequential incommensurability defined above, then we can now see that incommensurability does not make it impossible to assign to the two theories T_1, T_2 in question degrees of problem-solving power of any of its reasonable varieties and hence to compare their relative merits by means of their problem-solving power.

In other ways, too, we can see the validity of Laudan's main point, which is to criticize the equation of rational comparability with translatability. For one thing, the elimination of consequential incommensurability (say, from the relation of two theories T_1 and T_2 to each other) does not necessarily require enhancing their mutual translatability. One way of eliminating the consequential incommensurability of T_1 and T_2 is, not necessarily to strengthen either one, nor to introduce new translation rules to connect the two, but to improve the Inquirer's experimental techniques so that Nature will begin to answer more "small" questions than before and thereby make more "big" questions (that is, questions to be answered by means of the entire questioning procedure) answerable.[5] There are in

[5]The simplest version of an interrogative process starts (compare note 1 above) from a given theoretical premise T and aims at a fixed conclusion C. In a slightly more complex

fact many examples in the history of science where improved capacities of observation and experimentation have for the first time enabled scientists to compare the consequences of different competing theories. Such an "experimental" way of enhancing the commensurability of two theories would of course be incomprehensible if their incommensurability were merely a matter of conceptual alienation.

Of course, improved experimentation and observation is not the only way of enhancing the comparability (commensurability) of the two theories T_1 and T_2. Strengthening either theory T_1 or T_2 itself can likewise make them consequentially commensurable, which in a different way illuminates the fact that the lack of mutual translatability of T_1 and T_2 is not the only reason for the incommensurability of theories.

In the beginning of this paper, it was noted that the concept of incommensurability is not very useful for those philosophers who want to argue for the rational incomparability of theories unless this concept is given an independent characterization. Now it has been seen that, on what looks like the most natural explication of the concept of incommensurability, it does not offer any aid and comfort to defenders of incomparability. For on this explication the incommensurability of two theories does not imply their incomparability. What remains to be done is to strengthen this result further by developing my characterization of incommensurability further and by showing that it captures in fact the ideas it was intended to capture, especially the idea of conceptual discrepancy (untranslatability).

5. Measures of Incommensurability. A complete incommensurability of two theories is clearly only one end of a long spectrum of lower or higher degress of commensurability (comparability) in the sense discussed in the preceding sections. This observation prompts the question as to how the degree of commensurability of two given theories T_1 and T_2 is to be measured.

It was suggested above that the commensurability of T_1 and T_2 be measured by their shared consequences. This idea now has to be made more specific in several respects. First, the idea of consequence has to be understood so as to pertain to model consequences, not merely deductive consequences. Any model consequence C can always be considered as an answer to the question "C or not-C?". The natural idea in estimating commensurability is obviously to use the set of questions that both T_1 and

version, the aim of the game is to prove either C or $\neg C$, that is, to answer the question "C or not-C?". Then the inquirer must keep two scorecards, as it were, one for the potential conclusion C and the other for the potential conclusion $\neg C$. Needless to say, this initial "big" question has to be distinguished from the "small" questions by means of which the inquirer is trying to answer the initial one.

T_2 enable the Inquirer to answer as an index of their commensurability.

Second, we cannot consider these model consequences one by one in measuring commensurability, for they are not independent of each other. Instead, something like the conjunction of all the relevant consequences has to be considered.

Third, I shall first present a definition for the case in which T_1 and T_2 give the same answer to all the relevant questions that they both enable the Inquirer to answer. Later, I shall show how this definition can be generalized.

I shall relativize the definition of my measure of commensurability to a set Q of yes-or-no questions

$$Q = S_i \text{ or} \neg S_i? \quad (i \in I) \tag{5.1}$$

Let the set of answers to (5.1) that are yielded by T_1 be

$$(\neg)S_j \quad (j \in J_1) \tag{5.2}$$

and let the sets of answers yielded by T_2, by $(T_1 \cup T_2)$, and by both T_1 and T_2 separately, be likewise defined by index sets J_2, J_{12}, and $J_1 \cap J_2$, respectively.

The commensurability of T_1 and T_2 obviously depends on the information content of the shared answers. This information is

$$inf\,(\Lambda_{j \in J_1 \cap J_2}(\neg)S_j) \tag{5.3}$$

Its share of the information that the two theories yield jointly is

$$inf\,(\Lambda_{j \in J_1 \cap J_2}(\neg)S_j)/inf\,(\Lambda_{j \in J_{12}}(\neg)S_j) \tag{5.4}$$

This, clearly, is a natural measure of the commensurability designated by $Comm(T_1, T_2, M, Q)$. This definition highlights the fact that on my explication the commensurability of two theories is relative both to a model M and to a set of questions Q. In a special case, this set can be the set of all yes-or-no questions that can be formulated in the language of $T_1 \cup T_2$. In another interesting special case, it can be the set of all general yes-or-no questions that can be formulated in the language of $T_1 \cup T_2$, that is, that do not contain names of any members of M.

It is also possible and in fact useful to define a closely related asymmetrical measure of what might be called the commensurability of T_1 with respect to T_2. This can be defined as

$$inf\,(\Lambda_{j \in J_1 \cap J_2^{\prime}}(\neg)S_j)/inf(\Lambda_{j \in J_2}(\neg)S_j \tag{5.5}$$

(The usefulness of this notion is due to the fact that the interaction of the two theories can make it hard to determine J_{12} on the basis of T_1 and T_2.)

Furthermore, the notion of commensurability can be extended to sets of models Ω, on the assumption that a probability measure λ (a set function) is defined on the space of all models.[6] The measure will obviously be

$$\int_{\Omega} Comm(T_1, T_2, M, Q)\, d\lambda(M)/\lambda(\Omega) \tag{5.6}$$

This can be absolutized by choosing Ω = the set of all models of the language of $T_1 \cup T_2$.

These measures can be extended to the case in which the two theories T_1 and T_2 yield different answers to some of the questions which they both enable the Inquirer to answer. All we have to do is to replace (5.3) in (5.4) by the average of the following two expressions

$$inf\,(\Lambda_{j\in J_{01}} (\neg)S_j) \tag{5.7}$$

$$inf\,(\Lambda_{j\in J_{02}} (\neg)S_j) \tag{5.8}$$

where $(\neg)S_j$, $j \in J_{01}$, are all the answers that T_1 yields to the questions which both T_1 and T_2 enable us to answer and $(\neg)S_j$, $j \in J_{02}$ are the answers T_2 yields to the same questions. In (5.5), it is natural to replace (5.3) by (5.8).

Thus the consequential sense of commensurability admits of a simple and natural definition in terms of the interrogative model. It is worth noting that the measure of commensurability is relative to a given set Q of (yes-or-no) questions. Since the choice of this set can be influenced and even determined by extrascientific factors, my definition of commensurability in a sense is not a purely logical one, but allows the commensurability of theories to depend on such extrascientific factors as value-judgments. Likewise, the choice of the information measure used in (5.3)–(5.5) may be partly determined on extralogical considerations.

The definitions (5.4)–(5.5) of commensurability turn out to be closely related to what looks like the most straightforward explication of the notion of the problem-solving (question-answering) power of a theory (Hintikka 1984a, 1984b, 1985; Hintikka and Hintikka 1982). This connection further enhances the plausibility of my definition.

6. Example: The Classificatory Case. As an illustration of what has been said, it is instructive to consider the case of a language with only monadic predicates. In spite of the simplicity of this case, it serves to illustrate the merits of the definition of incommensurability offered in this paper.

If the primitive predicates of such a language are P_1x, P_2x, . . ., P_jx,

[6]For the logico-mathematical background of this extension, see, for example, Fenstad (1980).

then what can be done by their means is to define $2^j = J$ so-called q-predicates of the form

$$(\neg)P_1x \,\&\, (\neg)P_2x \,\&\, \ldots \,\&\, (\neg)P_jx \tag{6.1}$$

where "$(\neg)$" is to be replaced by "$\neg$" or omitted in all the different possible combinations in the different q-predicates. These q-predicates effect a partition of all (possible) individuals into pairwise incompatible and collectively exhaustive cells. A given individual belongs to precisely one such cell, that is, has precisely one q-predicate.

The strongest general sentences (that is, sentences not mentioning any particular individuals) are called *constituents*. Each constituent specifies which q-predicates are instantiated and which ones are not. There are obviously 2^J constituents. All general sentences are logically equivalent to disjunctions of constituents. Any two constituents (in the same language) are mutually exclusive. Hence constituents effect the finest partition of the space of all models with the same language, in the sense that in each model precisely one constituent is true.

In considering what happens to the concept of incommensurability in this monadic case we are of course dealing with the different theories T_1 and T_2 with partly different, partly overlapping vocabularies. Let us assume that the primitive predicates of T_1 are

$$P_1x, P_2x, \ldots, P_kx, Q_1x, Q_2x, \ldots, Q_mx \tag{6.2}$$

while those of T_2 are

$$P_1x, P_2x, \ldots, P_kx, R_1x, R_2x, \ldots, R_nx. \tag{6.3}$$

(The predicates Q_i and R_j are assumed to be different from each other.) The language of $T_1 \cup T_2$ consists of the union of (6.2) and (6.3). Any deductive consequence which T_1 and T_2 share must be formulated in the language T_1 and T_2 share, that is, in the language whose primitive predicates are

$$P_1x, P_2x, \ldots, P_kx \tag{6.4}$$

Let us suppose that we are considering interrogative consequences and not only deductive ones, and let us assume also that the only answers the inquirer can hope to receive are negated or unnegated atomic sentences. Furthermore, let the relevant questions (5.1) be those concerned with the truth or falsity of all closed sentences (that is, sentences free of names of particular individuals). Then the universal part of a constituent will be an optimal theory. For what such a part of a constituent says is which q-predicates *are not* instantiated. What the inquirer can find through the answers he or she receives is which q-predicates *are* instantiated. Hence by means of its model consequences (interrogative consequences) such a universal theory T_1 can be strengthened into the corresponding constituent C_1. Such an interrogatively optimal theory will be of the form

$$(\forall x) \bigvee_{i \in I} Ct_i[x] \tag{6.5}$$

where $\{Ct_i[x]\}$, $i \in I$ is a subset of the set of all q-predicates in the language of T_1.

I shall assume, by way of thought-experiment, that T_1 and T_2 are interrogatively optimal in this sense. Then each of them corresponds to a constituent, C_1 and C_2, respectively, where C_1 also says (over and above T_1) that all the q-predicates occurring in (6.5) are in fact instantiated. If T_1 is true in the "world" in which the interrogative inquiry takes place, then C_1 is the sum total of all the general conclusions that can be extracted from T_1 by the questioning procedure (assuming, of course, the restriction of answers to atomic sentences and their negations). The same holds of course of T_2. Let us call the number of predicates in C_1 (that is, the cardinality of the set I in (6.5)) c_1, and correspondingly for C_2 and c_2.

If the two constituents T_1 and T_2 are compatible, then the total information of answers obtainable from both T_1 and T_2 is the information of the constituent C_0 in the shared vocabulary of T_1 and T_2 which spells out what C_1 and C_2 say about this poorer set of concepts. Let us assume that according to C_0 there are c_0 cells instantiated definable in the shared vocabulary. In such circumstances, what is the degree of commensurability of T_2 with respect to T_1?

Consider first the information yielded by C_1. There are altogether 2^{K_1} constituents that can be formed in the language of T_1 (that is, where $K_1 = 2^{k+m}$). If they were all equally probable, each would carry the information amount

$$-log\,(1/2^{K_1}) = 2^{k+m}. \tag{6.6}$$

However, this is not the case, for the equiprobability distribution must be assumed to be with respect to constituents in the richer language of $T_1 \cup T_2$. In this richer langauge there corresponds to each of the c_1 cells instantiated according to C_1 precisely 2^n cells. For each cell we get

$$(2^{2^n} - 1) \tag{6.7}$$

different constituents in the richer vocabulary, hence altogether

$$(2^{2^n} - 1)^{c_1} \tag{6.8}$$

constituents in the pooled vocabulary. Hence

$$inf\,(C_1) = -log((2^{2^n} - 1)c_1/2^{2^{k+m+n}}) \tag{6.9}$$

which is approximately

$$2^{k+m+n} - c_1 \cdot 2^n \tag{6.10}$$

By the same token, we have, approximately,

$$inf\ (C_0) = 2^{k+m+n} - c_0 \cdot 2^{n+m}. \tag{6.11}$$

Hence (compare (5.5) above) the commensurability of T_2 with respect to T_1 is, approximately,

$$(2^{k+m+n} - c_0 \cdot 2^{n+m})/(2^{k+m+n} - c_1 \cdot 2^n)$$
$$= (1 - (c_0/2^k))/(1 - (c_1/2^{k+m})) \tag{6.12}$$

This is an eminently natural result. For instance, if $m = 0$, that is, if T_1 does not have any nonlogical constants not in the language of T_2, we obviously must have $c_0 = c_1$, whereupon (6.12) becomes one. Furthermore, (6.12) grows smaller when c_0 grows. This is indeed what the case ought to be.

Furthermore, a q-predicate in the shared vocabulary (6.4) that is implied by T_1 and by T_2 to be empty does not leave any further choices open. However, a similar q-predicate that is said to be instantiated leaves open all sorts of options as to which richer q-predicates compatible with the poorer one are instantiated. Hence a shared consequence concerning the positive instantiation of a q-predicate has a lesser weight in measuring the shared consequences of T_1 and T_2 than a negative result. Indeed, if $c_0 = 2^k$, that is, if all poorer q-predicates are instantiated, knowing this yields very little information as to which richer q-predicates are or are not instantiated, just as the (approximative) expression (6.12) shows.

What happens if T_1 and T_2 are not compatible in this example? It is easily seen that the derivation of (6.12) remains intact. However, now we may have to interpret C_0 as the constituent in the weaker (shared) vocabulary implied by C_1, which need not be the same constituent in the shared vocabulary that is implied by C_2.

The symmetrical measure of commensurability (5.4) will be somewhat more difficult to calculate in the case of incompatible theories T_1 and T_2 of the kind discussed, but does not yield essentially different conclusions from those suggested by (6.12). Since we are dealing here with an instructive example, anyway, there is no need here to study the more complicated measure.

7. Commensurability and Intertranslatability. The simple observations I have registered illustrate the naturalness of my definition of commensurability. They also open interesting general perspectives. Perhaps the most interesting one concerns the very nature of my definition of commensurability. It was motivated by the idea that the commensurability of two theories is measured by their shared consequences, in the sense of shared model consequences. This may seem to be a far cry from the prevalent idea of incommensurability as untranslatability. It was nevertheless pointed out informally in section 4 above that a definition of incommensurability that turns on shared model consequences need not fail to take into account conceptual discrepancies between the two theories that are being compared with each other. Now it can be seen that this informal point is in fact justified by an explicit analysis of the conceptual situation. In brief, my "consequential" definition of commensurability (5.6) does do justice also to the idea of incommensurability as (something like) untranslatability.

This can be seen both by means of general considerations and by means of illustrative examples. Using the latter strategy first, let us consider what happens when there are conceptual connections between the nonlogical primitives of the two theories T_1 and T_2. For instance, assume, for the sake of argument, that in the classificatory (monadic) example with optimal theories discussed in the preceding section, all the m nonshared predicates $Q_1, Q_2, \ldots, Q_m$ of T_1 are in fact definable in terms of the shared predicates $P_1, P_2, \ldots, P_k$ in the model M in question. Then the predicates $Q_1, Q_2, \ldots, Q_m$, being definable in terms of $P_1, P_2, \ldots, P_k$, will not cause any cells defined by q-predicates to divide when $Q_1, Q_2, \ldots, Q_m$ are added to the P's. This means, as can be seen from (6.12) and from the argument that led to it, that the commensurability of T_2 with respect to T_1 remains as it was, namely (6.12), while the commensurability of T_1 with respect to T_2 represented approximately by (6.12) with the roles of T_1 and T_2 exchanged becomes one. In other words, the definability of the nonshared predicates $Q_1, Q_2, \ldots, Q_m$ of T_1 in terms of the shared predicates $P_1, P_2, \ldots$ makes their effect on the incommensurability of T_1 with respect to T_2 disappear, thus increasing the commensurability of T_1 with respect to T_2.

In general, increasing the identifiability of the nonlogical vocabulary to T_1 or of T_2 in terms of that of the other tends to increase (*ceteris paribus*) the commensurability of T_1 and T_2 according to my definition, even though simple numerical results are not easily forthcoming in general.[7]

[7]Identifiability is in effect the interrogative-logic counterpart to definability (definability in each model). It is also a generalization of econometricians' namesake notion. A one-place predicate is identifiable in a theory T if and only if T implies a finite disjunction of "reference point" definitions of the form

$$(\exists x_1)(\exists x_2) \ldots (\exists x_k)(\forall y)(Py \longleftrightarrow D[x_1, x_2, \ldots x_k, y])$$

where the usual conditions on the *definiens* are satisfied. Compare works mentioned in note 1 above.

For instance, assume that T_1 is model-complete with respect to its own nonlogical vocabulary and that all the nonlogical constants of T_2 are identifiable on the basis of the vocabulary of T_1. Then it is easy to see that all the model consequences of T_2 are also model consequences of T_1, thus making T_2 optimally commensurable with respect to T_1.

Thus we can in different ways see that my definition of the degree of commensurability of two theories T_1 and T_2 serves also as an explication of the idea of commensurability as conceptual kinship not unlike translatability (though not identical with intertranslatability).

It is of interest to note that the bridge between the vocabularies of T_1 and T_2 which serves to enhance their commensurability need not consist in explicit or piecewise definability but can rely on identifiability instead.

The main moral of my story concerns the uses of the concept of incommensurability in recent philosophy of science. Even if we construe the incommensurability of two theories T_1, T_2 as conceptual discrepancy between them, the concept of incommensurability is not inaccessible to explicit logical analysis. Moreover, it turns out that even a complete incommensurability of two theories does not make them incomparable by various objective standards. For instance, they can be compared in terms of Laudan's concept of problem-solving power, which I will deal with in the forthcoming work mentioned in note 1 above. Hence the frequent arguments that strive to use the absolute or relative incommensurability of scientific theories as a reason for thinking that they are inaccessible to be purely scientific (rational) comparisons are simply fallacious.

REFERENCES

Fenstad, J.-E. (1980), "The Structure of Probabilities Defined on First-Order Languages", in Richard C. Jeffrey (ed.), *Studies in Inductive Logic and Probability,* vol. 11. Berkeley: University of California Press, pp. 251–262.

Hintikka, J. (1984a), "The Logic of Science as a Model-Oriented Logic", in Peter Asquith (ed.), *PSA 1984*. East Lansing, Mich.: Philosophy of Science Association, pp. 177–185.

———. (1984b), "Questioning as a Philosophical Method", in J. H. Fetzer (ed.), *Principles of Philosophical Reasoning*. Totowa, N. J.: Rowman & Allanheld, pp. 25–43.

———. (1985), "A Spectrum of Logics of Questioning", *Philosophica 35*: 135–150.

Hintikka, J., and Hintikka, M. (1982), "Sherlock Holmes Confronts Modern Logic: Towards a Theory of Information-Seeking By Questioning", in E. M. Barth and J. L. Martens (eds.), *Argumentation: Approaches to Theory Formation*. Amsterdam: John Benjamins, pp. 55–76.

Laudan, L. (1977), *Progress and Its Problems*. Berkeley: University of California Press.

12

THEORY-LADENNESS OF OBSERVATIONS AS AS TEST CASE OF KUHN'S APPROACH TO SCIENTIFIC INQUIRY

1. What is Kuhn up to?

The overall character of the ideas Thomas S. Kuhn has offered concerning the nature of scientific inquiry has been generally misunderstood, or, rather, misconstrued. (See Kuhn 1957, 1970.) Kuhn's ideas do not add up to a fully articulated analysis of the structure of the scientific process. Kuhn does not offer a theory of science which should be evaluated in the same way as, e.g., the hypothetico-deductive model of science or the inductivist one. What Kuhn does is best viewed as calling our attention to certain salient phenomena which a philosophical theorist of science must try to understand and to account for. We do injustice to Kuhn if we deal with his views as if they were finished products of philosophical theorizing. They are not. Rather, they are starting-points for such theorizing; they pose problems to be solved by a genuine theory of science.

If we do not realize and acknowledge this, we run the risk of attributing to Kuhn a singularly shallow philosophy of science. As he uses them, several of Kuhn's central concepts can scarcely accommodate the theoretical traffic they have been put to bear. For instance, in an earlier paper (Hintikka 1988a) I have shown that the notion of incommensurability of theories does not behave in the way Kuhn assumes. Of course the incommensurability of two theories goes together with their conceptual alienation from each other, but it also goes together with the discrepancy between their respective consequences and hence can be characterized by reference to the latter, contrary to what Kuhn clearly assumes.

The attitude of most philosophers to Kuhn's work is all the more surprising as Kuhn has himself sought a deeper analysis of, and a firmer theoretical foundation for, his ideas. For instance, at one point Kuhn (1977) took very seriously the possibility that the so-called structuralist approach of Stegmüller and Sneed might provide a satisfactory theoretical framework for his ideas.

This perspective is also relevant to the evaluation of Kuhn's argumentation. If you measure it against what one is entitled to expect of a philosopher with a command of the conceptual and structural issues involved in understanding scientific inquiry, you will find Kuhn's argumentation disappointing. For one thing, his argumentation is often, not to say typically, negative. Instead of developing his own views constructively, he presents his point by criticizing other views, irrespectively of whether any major philosopher has ever actually held them.

Furthermore, some of Kuhn's arguments come close to being self-refuting. For instance, Kuhn cites empirical studies from the psychology of perception to buttress his claim that our observations are influenced by background beliefs, including theoretical ones. But such evidence is a double-edged sword. For if a rule-governed influence of beliefs on observations has actually been established, our awareness of the very rules governing it can in principle be used to eliminate the effects of our theories from our observations, by compensating against the bias these theoretical beliefs induce.

Even Kuhn's historical argumentation is frequently unconvincing. For instance, Kuhn's own prize specimen of a scientific revolution, the Copernican revolution, is as good a counter-example to his thesis of the theory-ladenness of observations as one can hope to find. For the actual observations used by the rebels of the astronomical revolution were theory-neutral. They were sightings of heavenly bodies, data as to which heavenly body was where on the firmament when. Indeed, what made possible the discovery of Kepler's Laws were Tycho Brahe's observations, notwithstanding Brahe's rejection of the Copernican system. (Which theory were his observations laden with?) Copernicus' picture of the world changed profoundly the way we humans have to think of our place in the cosmos, but his argumentation in no way presupposes understanding or interpreting observational data in a new way. It is couched mostly in time-honored terms of simplicity and naturalness. If there is a new way of looking at the situation, it is to extend to the heavenly bodies the same dynamic questions as can be raised about terrestrial bodies. (See Grant 1962.) For instance, this parity of the terrestrial and of the heavenly, rather than the new Copernican picture as such, is what motivates the denial of the immutability of the heavens. (See Grant 1991.) But this new way of looking at things did not affect the role of observations in Copernicus' argumentation.

2. Theory-ladenness and the interrogative model of inquiry

Yet the questions Kuhn is raising are not only very real but also of considerable subtlety. The right way of approaching them is not to focus on what Kuhn says, but to try to put the phenomena he is calling our attention to into a deeper perspective. In this paper, I shall use Kuhn's idea of the theory-ladenness of observations as a test case. This idea is important for Kuhn, for without the theory-ladenness of observations several of his other central theses, for instance the frequent incommensurability of theories, would be considerably weakened.

At first sight it might seem that Kuhn's theory-ladenness thesis is correct but quite shallow, and cannot bear the demonstrative traffic Kuhn loads it with. It may be a welcome correction to the positivistic picture, which Kuhn is using as his strawman. But as soon as a more realistic picture of the scientific process is adopted, Kuhn's thesis can be accommodated without much ado. As an object lesson, I shall here consider the issue of theory-ladenness of observations in terms of the interrogative model of inquiry which I have developed and applied in the last several years. (Cf. Hintikka 1988b.) Here only the most general features of the model are needed. In the model, scientific inquiry is conceptualized as a questioning game between an inquirer and nature (more generally, any suitable source of answers generically referred to as an "oracle", or even several oracles). The inquiry starts from some given theoretical premises (or set of premises) T. In the simplest case, the aim of the game is to prove a given potential conclusion C. In doing so, the inquirer can perform two kinds of moves, interrogative and logical. A logical move is simply a logical inference from the results (premises) so far reached. In an interrogative move, the inquirer puts a question to nature. (The presupposition of the question must of course have been established prior to asking the question.) If an answer is forthcoming, it is added to the inquirer's list of available premises. It is assumed that the set of available answers is fixed, and that it remains constant throughout the inquiry.

The answers can be thought of (in the application of the model considered here) as including observations and results of experiments—any factual data a scientist can lay her or his hands on. In the simplest case, nature's answers are all assumed to be true, but this assumption can (and must) be given up in other, more complex types of inquiry. If the inquirer can derive C no matter what true answers nature gives, C is said to be interrogatively derivable from T in M, in symbols

(1) $M: T \vdash C$

where M is the model ("world") to which nature's answers pertain.

In the interrogative model, a scientist's observations can be construed as nature's answers (or as a subclass of nature's answers) to the inquirer's questions. They are among the premises that the inquirer can use in her or his logical inference steps, and they can also be used as presuppositions of further questions.

The basic fact about observations is that the role they play depends crucially on the rest of the inquiry. What can be done by means of a set of observations (for instance, what follows from them interrogatively) does not depend on these observations alone, but also heavily on theory the inquirer is relying on, that is, on his or her initial premises. If you change the initial premises, an observational answer will have entirely different interrogative consequences. This dependence of the role of observations on initial theoretical premises already amounts to a massive and clear-cut "theory-ladenness of observations".

Furthermore, in more realistic variants of the interrogative inquiry nature's answers are not assumed to be known to be all true. They can be accepted, rejected, reaccepted, and so forth, by the inquirer, in accordance with strategic rules which depend (over and above *a priori* knowledge about the probability of different answers' being true) on what happens in the rest of the interrogative inquiry, which in favorable circumstances can even be a self-correcting process. Among other things, these strategies depend on the given theoretical premises, thus revealing another kind of theory-dependence of observations in inquiry.

All this destroys the positivistic conception of observations as theory-neutral building-blocks more radically than Kuhn's arguments. At the same time, the interrogative model shows that the *theory*-ladenness of observations is not the only way they depend on their context in inquiry. The impact of an observation on inquiry does not only depend on the inquirer's initial theoretical premises. It normally also depends on the other answers the inquirer receives to her or his observational questions. This point should not come as a surprise to Kuhn. For instance, in discussing how earlier experience can affect our observations, Kuhn himself is not really talking about the theory-ladenness of observations. He is in effect talking about the observation-ladenness of observations.

The range of answers which nature will give often is an even stronger determinant of the role of an observation in interrogative inquiry than the set of its initial theoretical premises. In this sense, too, observations are thoroughly laden with (the totality of available) observations. This fact is one of the many reasons why I said above that on its *prima facie* construal Kuhn's theory-ladenness thesis is relatively shallow.

3. Interim conclusions

Thus in a sense the interrogative model vindicates splendidly Kuhn's thesis of the theory-ladenness of observations. In doing so, however, it puts Kuhn's ideas in a new light. For one important thing, we can now see that theory-ladenness does not have to be explained in the way Kuhn and others have tried to explain it, to wit, as being due to the dependence of the meanings of observational terms on the theory in which they occur. From the fact that the *meaning* (in the sense of significance) of an observation for a scientific inquiry depends on the theory on which this inquiry is based, one can only at the risk of the fallacy of equivocation infer that the observation *means* something different in different theories in any sense of "meaning" related to logical or linguistic meaning, e.g., in the sense that the linguistic meaning of observational terms depends on the theory in which they occur. Theory-ladenness of observations simply does not presuppose that the meaning of observational terms changes from theory to theory or from world to world. The way in which the interrogative model vindicates the idea of theory-ladenness is in many ways faithful to the ideas of the likes of Kuhn and Hanson, but it does not involve any dependence in the usual sense of the meanings of observational terms on the theory in which they occur.

Admittedly, there exist philosophical views on linguistic meaning which tend to make it theory-dependent. Frequently, however, such views are defended on the basis of ideas attributed to Kuhn, and hence cannot serve to defend Kuhn. And even on the most favorable perspective on their independent justification, the jury is still out on their justification.

In a different direction, the theory-ladenness of observations in the sense uncovered by the interrogative model does not automatically entail any unavoidable incommensurability of theories, either. (See Hintikka 1988a.) This incommensurability was thought to be a corollary to the dependence of the meaning of observational terms on the background theory, but (as was seen) there is no need whatsoever to assume the confused and confusing idea of meaning relativity in the first place. Theories can, of course, be more or less incommensurable also on the interrogative model, but this merely means that they have different (interrogative) consequences that can be tested against nature's answers. It even turns out, at least in oversimplified but representative cases, that this kind of incommensurability-without-meaning-variation is reconcilable with the idea that the incommensurability of two theories is due to conceptual discrepancies between them. Indeed, the observational incommensurability can be shown to be the smaller, the more conceptual ties there are between the terms occurring in the two theories in question. Hence once again we do not need meaning variation from theory to theory in order to do justice to the interesting phenomena to which Kuhn has called our attention.

4. On the logic of experimental science

Thus we seem to have reached a comfortable and comforting conclusion about what is true and what is false about Kuhn's claims. However, what is even more interesting than the conclusions so far reached is the possibility of pushing our analysis deeper. I suspect that Kuhn also saw deeper into the situation than I have so far brought out, even though he did not have the conceptual tools to articulate his insights. It turns out that theories are indeed involved in observations in a more basic way than we have diagnosed so far. It is not only that the consequences (and the other repercussions) of an observation are arguably theory-dependent. The very possibility of an observation may be contingent on theoretical assumptions.

In order to see what's what here, let us first register a *prima facie* objection to what I have said so far. It might seem that Kuhn could very well claim that the interrogative model tells only a part of the full story. The kind of inquiry the model codifies seems at first sight to match very well what Kuhn calls normal science. Indeed, Kuhn characterizes normal science as an exercise in puzzle-solving. In contrast, the interrogative model might not seem capable of handling the situations in which a revolutionary change takes place in a science or even, on a more modest scale, the kind of situation in which an entire new theory replaces an old one.

Nevertheless, it is precisely in the problem area of theory change that the interrogative model begins to show its real strength. A number of further insights will enable us to see what's what.

First, why is the interrogative model supposed to be incapable of serving as a paradigm for an inquiry which establishes a radically new theory, perhaps even without presupposing any strong initial premises? Here we come to an assumption which is weighty, widespread and wrong. It amounts to assuming that the only answers nature can give an inquirer are particular propositions. I have called this assumption the Atomistic Postulate. I have argued that it lies behind much of the traditional philosophy of science. Here I am suggesting that it also underlies Kuhn's thinking. It is because of this assumption that philosophers have thought that scientists cannot literally derive their theories from nature's answers without strong initial theoretical assumptions. I have also shown that this postulate is wrong in the sense that it does not adequately capture the structure of actual scientific reasoning. For instance, the answers nature gives in a successful controlled experiment express dependencies between two variables and hence has at least the quantificational complexity of a $(\forall x)(\exists y)$ prenex. More complex experimental setups can in principle yield even more complicated answers. (See Hintikka 1988b.)

From such answers the inquirer can logically infer complex laws and theories even without the help of strong initial theoretical premises and without the help of any inductive or other ampliative rules of inference. Even though I will not pursue the point any further here, this rejection of the Atomistic Postulate puts large segments of the philosophy of science to a new light.

5. Scientific inquiry as a two-level process

But even this analysis of the logic of experimental science is not enough to do full justice to the nature of experiments. For even though the results of an experiment typically enter into the reasoning in the form of dependence laws, the way in which those laws are reached must be capable of being analyzed.

The natural solution here is to consider scientific inquiry in theoretically sophisticated sciences as a two-level process. This two-level perspective is made possible by the double role of questions in interrogative inquiry. In more complex types of inquiry, the inquirer's goal is not to prove a predetermined conclusion (for example, to verify a hypothesis) but to answer a question. How the techniques of interrogative inquiry can be used to this end is not an easy question to answer. An answer covering all cases can nevertheless be given. In general, questions play two roles in interrogative inquiry. What happens is that the inquirer tries to answer a "big" (principal) question by means of a number of "small" (operational) questions. In any one inquiry, the two questions have to be distinguished from each other sharply. However, what for a higher-level inquiry is an operational ("small") question can for the purposes of a lower-level inquiry be the principal question of a complex inquiry in which it is to be answered by means of a number of lower-level operational questions. This, I find, is how we must view typical controlled experiments. For the purposes of a higher-level inquiry, the entire functional dependence (of the observed variable on the controlled variable) that is the outcome of the experiment is an answer to an operational question on the higher (theoretical) level. For the experimental scientist, in contrast, it is an answer to a principal question, and the experimentalist's operational answers are particular data brought to light during the experiment, for instance, instrument registrations.

What is interesting here is not just that experimental inquiry can be considered as a two-level process, but that the two levels have distinctly and characteristically different structures. The operational answers on the higher level can have a considerable logical complexity, whereas on the lower-level operational answers are typically formulated as particular propositions. Even more interestingly, the higher-level inquiry does not necessarily need strong initial premises, whereas the particularity of the lower-level operational answers presupposes that suitable general premises are available to the (experimental) inquirer.

Where do these general premises come from? The approximate answer is: The initial premises of a lower-level inquiry are theoretical laws established earlier on the higher level. Typically, they are not the ones that are being investigated or tested on the higher level; they are older and safer (and frequently only partial) generalizations.

I cannot argue here fully for this reading of the typical situation, but it should not be news to anyone familiar with experimental techniques in sciences like physics. It is also supported, it seems to me, by the recent studies of the role experiments in science by the likes of Alan Franklin (1986, 1990) and Peter Galison (1987).

If the view I have presented of the two-level character of scientific inquiry is right, it shows in some real detail how it is that an experimental inquiry depends on general laws and even on theories binding such laws together. In brief, it provides an account of what might be called the *theory-ladenness of experiments*. This account does not serve merely cosmetic purposes, either. It shows among other things what role induction does and does not play in actual scientific inquiry. On the higher level, induction is not needed, because the operational answers can have such a logical complexity that laws and even theories can in principle be derived from them deductively. The place of inductive inference is taken over by the kind of reasoning which is involved in extending a partial generalization and combining different partial generalizations with each other into a more sweeping generalization. As it happens, such reasoning was earlier in the history known under the very label "induction". (See Hintikka 1992.)

On the lower level, induction is needed because operational answers are usually particular propositions. But on this level the requisite general premises needed to back up so-called inductive inference are normally available, supplied by the results of earlier higher-level inquiry. Thus the (modest) role of induction in actual science receives a diagnosis in this way.

The two-level model can also be compared with the testimony of the actual history of science. Even though all the returns are not yet in, there already are several encouraging early results. For one important example, Newton's methodological pronouncements and methodological practice are not only put to a new light which strikingly vindicates some of them, for instance, Newton's claim of having "deduced" general truths from "phenomena"; what is more, Newton's theory and practice of methodology are found to be in an excellent conformity with each other (and with the interrogative model), current orthodoxy notwithstanding. (See here Hintikka and Garrison, forthcoming.)

6. The parallelism of experiments and observations

But how is this analysis of experiments relevant to the theme of this paper? I have argued for the theory-ladenness of *experiments*, but what does that have to do with the vaunted theory-ladenness of observations? The answer lies in a parallelism which I see there between experiments and observations. Such a parallelism seems to be to be implicit in Kuhn's and Hanson's thinking, and it can in any case be argued for in the same way as they argue for their views. From classics like von Helmholtz to contemporaries like Rock, most of the sophisticated psychologists of perception have recognized that unconscious cognitive processes, variously described as "unconscious inferences" (von Helmholtz), "a kind of problem solving" (Gregory, 1970, p. 31) or "hypothesis testing" (Rock, 1983), play an important role in perception. Assimilating these processes to experiments is merely to put a generic name to what these sundry descriptions have been attempts to capture. In so far as the parallelism between an experiment and an observation is an apt one, the theory-ladenness of experiments which we have discovered serves as an explanation also for the theory-ladenness of observations.

7. On the two-levelled character of experimental inquiry

The two-levelled character of experimental inquiry has interesting consequences even apart from the problem of the theory-ladenness of observations. Here I can mention only a couple of perspectives it opens. Clark Glymour has called our attention to the feature of scientific inquiry he calls "bootstrapping", which he describes as follows:

> ... the basic idea is clear enough: Hypotheses are tested and confirmed by producing instances of them; to produce instances of theoretical hypotheses we must use other theoretical relations to determine values of theoretical quantities; these other theoretical relations are tested then in the same way. (Glymour 1980, p.52)

Such explanations leave a multitude of questions unanswered. The problem is not just that there does not seem to be any guarantee that the process is free of circularity and that it therefore is likelier to succeed than the attempt by Baron von Münchausen which presumably lent Glymour's idea (attributed by him to Reichenbach) its name. The disturbing question is how there could be two different sets of theoretical hypotheses governing precisely the same phenomena. How can a blind hypothesis lead another blind hypothesis?

The most straightforward answer is to view the process which Glymour assumes can produce instances of theoretical hypotheses (hypotheses$_1$) by means of other hypotheses (hypotheses$_2$) as a lower-level experimental inquiry carried out to test hypotheses$_1$. Then hypotheses$_2$ are the theoretical assumptions on which the lower-level experimental inquiry is based. This perspective is obviously much more faithful to actual scientific practice than, e.g., a conventionalistic attempt to avoid the specter of circularity.

One can even raise the question as to what happens if the two sets of hypotheses or theories are incompatible. It has been claimed that this is the situation we encounter in quantum-theoretical experiments where the experimental situation apparently must be dealt with classically even though the target phenomena are quantum-theoretical. The perspective we have reached here promises new possibilities of conceptualizing such questions.

8. Theory-ladenness and the logic of identification

Another direction in which we can analyze the situation further relates to the alleged dependence of the meanings of observational terms on the underlying theory of the inquirer. Here, once again, Kuhn's argumentative strategy is unclear. For what is the meaning of the "meaning" which is supposed to vary from theory to theory? What is uncontroversially true is that when different theories are true in two different "possible worlds" (under two different possible courses of events), a term will apply to different cases in the two worlds. But in the normal understanding of the semantics of our language, this can be the case even if the meaning of the term does not vary from one world to another in the slightest. And if in some sense its meaning is different in the two scenarios, why should Kuhn or anyone else speak of the variation of the meaning of one and the same term, instead of two different terms? Clearly the issues have to be sharpened here before they can be resolved.

The most straightforward way of interpreting Kuhn is to take him to claim that the dependence of the reference (extension) of a term on the theory in which it occurs as well as on the world which is being investigated must be explained by saying that the meaning of that term changes when the theory is changed. In order to see whether such claims are true, we have to examine precisely *how* the extension (range of correct applications) of a term, say P, depends on the underlying theory, say T[P].

Once again, the interrogative model proves its mettle in answering this question. A good testing-ground is provided by the question: When does a theory T[P] containing a term, say a one-place predicate P, determine completely the range of correct applications of P? Here the interrogative model opens an interesting perspective in that it shows that this answer admits of two different kinds of answer. The difference between them is related to philosophers' and linguists' much abused distinction between *de dicto* and *de re* statements.

Speaking *de dicto*, the natural explication of the determinacy question is to ask: Does T[P] have among its interrogative consequences a quasi-definition of the following form:

$$(2) \qquad (\forall x)(Px \leftrightarrow D[x,a_1,a_2, ..., a_k])$$

Here $D[x,a_1,a_2, ..., a_k]$ must satisfy the usual requirements of a definiens. Furthermore, $a_1, a_2, ..., a_k$ are members of the domain of individuals of the model to which the inquiry pertains. (Naturally, (2) must be interrogatively derivable without answers containing P.) If the answer is yes, i.e., if (2) is so derivable, P is said to be *identifiable de dicto* in M.

The question whether P is identifiable in M depends of course on T[P], and so does the available quasi-definition (2). In this sense, the reference (extension) of P is indeed theory-laden. This observation is trivial, however, and can be considered an explication of the comment above to the effect that the reference of a term depends on the theory in which it occurs. The real question is: What else can be said here?

One interesting result here is the following: If P is *identifiable de dicto* (in the sense explained above) in every model of the theory T[P], then it is piecewise explicitly *definable* on the basis of T[P], that is, there is a disjunction of explicit definitions

(3) $\vee_i (\forall x)(Px \leftrightarrow D_i[x])$

logically implied by T[P]. This result follows from a well-known theorem of Svenonius. (See Rantala 1977, p.79).

What this observation means is clear. By the meaning-determination of a term like P by the theory T[P] in which it occurs we must of course mean a determination applicable to all models of T[P]. The differences between different models are due to contingent facts independent of T[P]. But our result shows that the only way in which a theory can determine the reference of one of the terms it contains for all its models —which presumably is the sort of meaning determination at issue here—is by way of logical consequence.

In other words, not all kinds of theory-ladenness of meaning make any philosophical difference. Complete meaning determination *de dicto* is possible only in the old-fashioned way, to wit, by way of logical consequence.

However, these consequences follow only if identifiability is construed *de dicto*. If it is understood *de re*, then P is identifiable on the basis of T[P] in a model M iff for each member b of the domain of individuals do(M) of M, b ε do(M), we have either

$$M{:}T[P] \vdash Pb \quad \text{or}$$
$$M{:}T[P] \vdash {\sim}Pb\,.$$

It can be seen that P can be identifiable *de re* in M and yet not captured by any definitory formula like (2). Then Svenonius' theorem does not apply, and even a complete determination of the extension of a concept by a theory, viz. in the sense of *de re* identifiability, does not reduce to logical consequence.

This throws some light on the subtleties of meaning determination by theories. But independently of all these subtleties, and independently of how we prefer to express ourselves, one thing is clear. Neither in the case of the *de dicto* identifiability, nor in the case of *de re* identifiability, does the dependence of the range of correct application of a term on the theory in which it occurs have absolutely anything to do with the relativity of the meaning of the term to the theory. On the contrary, this kind of meaning dependence is obscured it we do not represent the concept expressed by the term in different ways in different theories or think that statements made in its terms on the basis of different theories are incommensurable for that would make it impossible to talk about logical consequence relations here. There is a solid truth in the claim of the theory-ladenness of meaning, but this point is to be demystified and brought to the purview of sober logical analysis. In particular, it would be an extremely serious mistake to think that because of the theory-ladenness of meaning somehow our usual logic has to be suspended or modified.

References

Earman, J. (1977), "Theory-Change as Structure Change", in *Historical and Philosophical Dimensions of Logic, Methodology and Philosophy of Science*, R.E. Butts and J. Hintikka (eds.). Dordrecht: D. Reidel, pp. 289-309.

Franklin, A. (1986), *The Neglect of Experiment*. Cambridge: Cambridge University Press.

_______. (1990), *Experiment, Right or Wrong*. Cambridge: Cambridge University Press.

Galison, P. (1987), *How Experiments End*. Chicago: University of Chicago Press.

Glymour, C. (1980), *Theory and Evidence*. Princeton: Princeton University Press.

Gooding, D. (1990), *Experiment and the Making of Meaning*. Dordrecht: Kluwer Academic.

Grant, E. (1962), "Late Medieval Thought, Copernicus and the Scientific Revolution", *Journal of the History of Ideas* 23: 197-220.

______. (1991), "Celestial Incorruptibility in Medieval Cosmology 1200-1687", in *Physics, Cosmology and Astronomy 1300-1700*, S. Unguru, (ed.). Dordrecht: Kluwer Academic, pp. 101-27.

Gregory, R.L. (1970), *The Intelligent Eye*. London: Weidenfeld & Nicolson.

Hacking, I. (1981), *Scientific Revolutions*. Oxford: Oxford University Press.

Hanson, N.R. (1958), *Patterns of Discovery*. Cambridge: Cambridge University Press.

Hintikka, J. (1988a), "On the Incommensurability of Theories", *Philosophy of Science* 55: 25-38.

_______. (1988b), "What is the Logic of Experimental Inquiry?", *Synthese* 74: 173-88.

_______. (1991), "Toward a General Theory of Identifiability", in *Definitions and Definability*, J.H. Fetzer, D. Shatz and G. Schlesinger, (eds.) . Dordrecht: Kluwer Academic, pp. 161-83.

_______. (1992), "The Concept of Induction in the Light of the Interrogative Approach to Inquiry", in *Inference, Explanation and Other Philosophical Frustrations*, John Earman (ed.). Berkeley and Los Angeles: University of California Press (forthcoming).

_______. and Garrison,J.W. (forthcoming), "Newton's Methodology and the Interrogative Logic of Experimental Inquiry", in *Science in Context*.

Kuhn, T.S. (1957), *The Copernican Revolution*. Cambridge: Harvard University Press.

_______. (1970), *The Structure of Scientific Revolutions*. 2nd ed., enlarged. Chicago: University of Chicago Press.

Rantala, V. (1977), *Aspects of Definability*. Helsinki: Societas Philosophica Fennica.

Rock, I. (1983), *The Logic of Perception*. Cambridge: MIT Press.

13

RAMSEY SENTENCES AND THE MEANING OF QUANTIFIERS

1. RAMSEY SENTENCES AND THE FUNCTION OF THEORETICAL CONCEPTS

In his famous paper "Theories" Frank Ramsey (1931, pp. 212-236; 1978, ch. 4) introduced a technique of examining a scientific theory by means of certain propositions, dubbed later "Ramsey Sentences." They are the results of what is often called Ramsey elimination. This *prima facie* elimination is often presented as a method of dispensing with theoretical concepts in scientific theorizing. The idea is this: Assume that we are given a finitely axiomatized scientific theory

(1) $T[O_1, O_2, ..., H_1, H_2, ...]$

where O_1, O_2, .. are the primitive observation terms (individual constants predicate constants, function constants, etc.) of (1) and H_1, H_2, ... its primitive theoretical terms. For simplicity, it will be assumed that (1) is a first-order theory. Since it is finitely axiomatizable, we may think of it as having the form of a single complex proposition, i.e. the conjunction of all the axioms. What can then be done is to generalize existentially with respect to (1). The result is a sentence of the form

(2) $(\exists X_1)(\exists X_2) ... T[O_1, O_2, ..., X_1, X_2, ...]$

In (2), the theoretical terms H_1, H_2, ... do not occur any longer. They have been replaced by the variables X_1, X_2, ... bound to initial existential quantifiers. In this sense at least, Ramsey sentences do effect on elimination. Unlike (1), (2) is not a first-order sentence but a second-order one.

The sentence (2) is known as the *Ramsey sentence* or *Ramsey reduct* of (1). As Ramsey was the first one to point out, all the logical consequences of (1) containing only the observational terms O_1, O_2, ... are also consequences of (2). Because of this, and because of the absence of any primitive theoretical constants from it, (2) is often thought of as being a genuine reduct of (1), that is, the result of eliminating all the theoretical concepts from (1). And, admittedly it does look as if we can be perfectly happy with (2) instead of (1) as long as we are only interested in the observational consequences of (1).

This idea of Ramsey sentences has provoked an extensive literature. Most of the discussion is inconclusive, however. Here I will consider other papers on Ramsey elimination only insofar as they impinge on my own line of thought.

The first main point to be made here is that Ramsey sentences have a great deal of interest completely apart from any questions of eliminability of theoretical terms. In fact there is a more fundamental question: How do the theoretical terms contribute to the import of a theory? This question was in effect raised already by Ramsey. And, having the exquisite sense for structural ("model-theoretical") aspects of logic and mathematics that he had, he correctly diagnosed the way theoretical concepts of a theory, say (1), can contribute to the import of (1) even as far as observational concepts are concerned. In order to see this, let

(3) $M(O_1, O_2, ..., H_1, H_2, ...)$

be a model of (1). If we disregard the theoretical concepts altogether, we obtain a substructure of (3) consisting of the observational concepts only. I will call it

(4) $M(O_1, O_2, ...)$

What we are inquiring into here is the input of the theoretical concepts into what the theory (1) says about the structure (4) characterized in the sole terms of the observational terms, in other words, how they restrict the class of models (4).

If the question is put in this way, the answer is clear. The theoretical concepts of (1) restrict the class of observational models (4) indirectly, without saying anything about observational concepts. They do so by requiring that each model (4) be imbeddable in a model (3) of the original theory (1). This is a purely structural requirement, as far as theoretical concepts are concerned, but it nonetheless is a very real one. The imbeddability requirement is the way in which the theoretical terms of a theory contribute even to its observational component, not through some tacit definitions or bridge laws but through the imbeddability requirement. And, as Ramsey clearly saw also, this can be a nontrivial contribution, even though it is a purely structural one. The fact that theoretical concepts obviously impose only purely structural requirements on the models of a theory like (1) need not make their contribution trivial in the least.

The way Ramsey highlighted this insight is by reference to Ramsey sentences. As long as we do not have any collateral information about the theoretical concepts, the imbeddability requirement is purely structural. It does not matter which particular individuals (or n-tuples of individuals) theoretical concepts apply to, as long as the structure they form (in relation to the observational concepts) satisfies (1). This indifference is shown already by the form of Ramsey sentences. They are existential statements. Hence it does not matter for the imbeddability requirement what the theoretical concepts $H_1, H_2, ...$ are, in the sense of which individuals (and n-tuples of individuals) they apply to. Therefore the Ramsey sentence (2) of the theory (1) carries precisely the same imbeddability requirement as the given theory. In other words, the Ramsey sentence says the same thing about the observables as the original given theory whose Ramsey sentence it is.

These are the insights that Ramsey emphasizes by his analogy of a Ramsey sentence and a fairy tale. It does not matter so to speak where the theoretical concepts are realized in the world, as long as they are realized somewhere, even if only in a fairy tale, so to speak. There is a metaphysical joke which I first heard from Marx Wartofsky, about the wise man who was convinced that God is such an absolutely perfect creature that He does not have to exist to exert His influence. Theoretical concepts are less perfect than God. They have to exist in order to do their job. But they come close, for their mere existence suffices to give them the powers they have, as far as observational concepts are concerned.

An important consequence of these observations is that in Ramsey sentences no role is played by tacit definitions or other bridge principles connecting theoretical and observational concepts with each other, unless they are included in the given theory. On the contrary, the crucial idea is that the given theory itself is the only bridge principle.

For this reason, it is inappropriate to give the title "Ramsey sentence" only to what in my perspective is the result of adding certain bridge sentences to (2) as some philosophers have suggested doing. This seems to be in agreement with Ramsey's intentions. Either the bridge sentences are a part of the original theory, and hence nothing new in (2), or else they are added in the "Ramsey elimination," i.e. in the step from (1) to (2). But in the latter case Ramsey could not maintain that a switch from (1) to (2) does not affect a scientist's reasoning.

The purely structural character of the job theoretical concepts do which is revealed by these observations relates the study of Ramsey sentences to the views according to which all that science can do is to study certain structures. It also poses the question as to what logical tools are the most appropriate ones for the study of such purely structural conditions as are imposed by theoretical concepts on theories. I will return to this question below.

2. RAMSEY SENTENCES DO NOT EFFECT AN ELIMINATION OF THEORETICAL CONCEPTS

But in what sense do Ramsey sentences effect an elimination of theoretical terms? Ramsey himself asserted that

> we can easily eliminate the functions of the second system and so say in the primary system all that the theory gives us. (1931, p. 219; 1978, p. 108)

Here "second system" means a theory involving theoretical concepts and "primary system" a theory containing only observational concepts ("functions," in Ramsey's terminology).

But the step from a theory like (1) to its Ramsey sentence (2) involves more than merely moving from a theory with theoretical terms to one without them. It also involves the introduction of higher-order quantifiers. If the given theory (1) is first-order, then the Ramsey reduct (2) is a *second-order* sentence. For this reason, a step from (1) to (2) cannot be simply identified with a switch from "system II" to "system I" in Ramsey's sense.

Speaking generally, it seems to me that one of the main sources of confusion concerning Ramsey sentences is an insufficient sensitivity on the part of philosophers of science to the differences between first-order and higher-order logic. This may be understandable in the case of Ramsey who was writing at a time when first-order logic had barely been formulated as an independent system (by Hilbert and Ackermann in 1928) and who had grown up on the tradition of *Principia Mathematica* with its higher-order logic (cf. here Moore 1988). However, the same excuse cannot be offered for his successors.

At first sight, the alleged Ramsey elimination does not change the situation at all. It seems to me that the fashionable talk about Ramsey *reduction* is due largely to thinking about propositions like (1) and (2) purely linguistically (formally). If we think of the situation model-theoretically (semantically), the picture looks quite different. True, (2) does not contain any theoretical constants. But the values of those constants must be among the values of the second-order variables X_1, X_2, ... in (2), for otherwise existential generalization is not valid. But if "to be is to be is to be a value of a bound variable," then (2) presupposes the existence of those very concepts which were thought of as being dispensed with in Ramsey elimination.[1] Indeed, (2) is naturally understood as asserting their very existence. Hence the step from (1) to (2) is not an elimination semantically or ontologically. As long as we quantify over second-order predicates, which inevitably include theoretical ones, we are assuming their existence, and no substantial reduction can therefore be involved.

Thus it is misleading to think of Ramsey sentences as carrying out a genuine elimination of theoretical concepts. This illusion is due merely to far too narrow a notion of how different concepts (in this case, theoretical concepts) can enter into scientific discourse. They do not have to be imported into a language by constants; they can also be smuggled in in the form of values of higher-order quantifiers.

These observations have several interesting consequences. For one thing, they imply that the possibility of Ramsey elimination has no bearing on the issue of realism vs. instrumentalism. Or, if it has, its impact is diametrically contradictory to what it is often taken to be. Ramsey reduction does not give any aid and comfort to the view that theoretical concepts "are simply language tools for organizing the observational phenomena of experience into some sort of pattern that will function efficiently in predicting new observables" (Carnap 1966, p. 255). On the contrary a Ramsey sentence is the bluntest possible way we have of saying that the theoretical concepts of a scientific theory really exist, viz. by means of existential quantifiers. The reality of this existence is very much involved in the way they operate. They form, together with the observables, a structure into which the observational part of the structure must be imbeddable. If this structure which includes the theoretical concepts were not real, the imbeddability requirement would be vacuous. The fact that we cannot locate the theoretical part of this structure in some particular nook or corner of the world is only another way of saying that we are dealing with theoretical concepts.

I have to push my line of thought further, however, and even challenge the interim conclusions so far reached. The comforting conclusions I just registered are not the last word on the subject.

They are not new, either. The most careful writers on the subject generally agree that what I have called Ramsey sentences do not implement a genuine reduction. For instance, Sneed not only notes that the function of theoretical terms lies in the imbeddability requirement (1971, p. 52) but defines a stronger notion of eliminability which he calls simply "Ramsey eliminability" (p. 53). Its motivation can be seen from what has been said in this paper. If second-order quantifiers involve ontological commitments to theoretical concepts, then we have a true elimination only if they too can be dispensed with. In this sense, which I will call (unlike Sneed) *strong Ramsey eliminability*, the theoretical concepts $H_1, H_2, \ldots$ are eliminable from (1) if and only if there is an ordinary first-order sentence

(5) $R[Q_1, O_2, \ldots]$

logically equivalent with (2).

As Sneed points out by means of examples (p. 54), theoretical concepts are not always strongly Ramsey-eliminable. Sneed does not try to characterize the difference between cases where strong Ramsey eliminability holds and those in which it does not hold.

This failure of strong Ramsey eliminability is an intriguing phenomenon. Somehow theoretical concepts can bring in additional force into one's theory even when they figure these only as values of second-order variables. How is this possible?

This question looks puzzling only when considered in terms of axiom systems with their axioms and rules of inference as a medium of deductive derivation of consequences from axioms. When the problem is considered model-theoretically, however, the situation becomes clearer.

An answer is in fact implicit in what was said earlier. The contribution of theoretical concepts to a given theory (1) was seen to lie in the requirement that, roughly speaking, the observational part (4) of any model (3) of (1) be imbeddable in some model of the form (3). But (3) goes beyond (4) only with respect to theoretical concepts, which by definition cannot be located in this or that part of the world. But their very existence imposes a condition on (4) that can be nontrivial.

For this reason, Ramsey elimination does not change this picture in the least. The theoretical concepts $H_1, H_2, \ldots$ as they occur in the given theory (1) impose through (1) definite conditions on the observable structures $M(O_1, O_2, \ldots)$. And this they do in the same way independently of whether the theoretical concepts figuring in (1) are introduced as values of nonlogical constants or as values of second-order quantifiers. In order to do their job, theoretical concepts merely need to exist in certain relations to the observational concepts of the theory, but it does not matter how or where they exist. In such a model-theoretical perspective, the Ramsey reduction does not change anything at all. It merely means that the left hand (predicate constants) lends money (condition-imposing power) to the right hand (second-order quantifiers).

It is important to realize what the purely structural requirement imposed on the models of observational concepts $M(O_1, O_2, ...)$ by the contribution of the theoretical concepts in (1) or of the second-order quantifiers in (2). The requirement is the same in the two cases. It is the imbeddability of $M(O_1, O_2, ...)$ in a richer structure $M(O_1, O_2, ..., X_1, X_2, ...)$ which is a model of (1). The force of this requirement does not depend on the presence of any definitions or other bridges between theoretical and observational concepts (besides theory (1) itself).

3. RAMSEY SENTENCES AND THE NOTION OF STRUCTURE

So what is the point of a Ramsey "elimination?" In order to see what is going on, let us consider two related objections to the idea of Ramsey elimination, as it was just presented. First, it might be objected that a Ramsey elimination is nontrivial only if a nonstandard interpretation of quantifiers is adopted. In other words, when it is asserted by the initial existential quantifiers of a Ramsey sentence, more is meant (so the suggestion goes) is than just the possibility in principle of specifying certain extensions (classes, classes of pairs, etc.) that satisfy the matrix of the Ramsey sentence. What is meant (so the claim goes) that there somehow exist in the nature of things themselves certain natures or forms that satisfy the matrix of the Ramsey sentence. Logicians know this as a nonstandard interpretation of higher-order variables as distinguished from the purely extensional "standard" interpretation. (For the distinction and for Ramsey's grasp of it, see Hintikka 1995(a).)

Alas, not only does Ramsey (1931, p. 220; 1978, p. 109) virtually in so many words opt for a purely extensional ("standard") interpretation of his language, (cf. his disjunctive definitions) what is more, the interest of the Ramsey elimination is in my judgment tied to the standard interpretation.

Second, it has been objected that Ramsey elimination is inevitably trivial. This objection can be illustrated by considering as a thought-experiment the elimination of *all* primitive predicates, observational as well as theoretical, from the given theory (1) in favor of variables bound to initial existential quantifiers. Such an elimination means that instead of describing the world by means of a number of concepts, expressed in language by interpreted constant predicates, including relational predicates, we merely describe the structure that the given theory imposes on the world. According to some philosophers, this is all that a theory can do in any case.

But are such views viable? In order to find an answer, we have to ask: What conditions does such a super-Ramsey sentence impose on its models? Of course, this is merely another way of asking: What does the super-Ramsey sentence say about reality?

The answer might at first sight seem paradoxical. The super-Ramsey sentence says that a structure of a certain kind can (purely extensionally speaking) be defined on the domain of individuals of the model in question. And, as is easy to see, this possibility depends only on the cardinality of the domain. Hence the only condition that a super-Ramsey sentence can impose on the world is some restriction of the cardinality of its denizens. Super-Ramsey sentences are in this sense completely trivial as a means of theorizing about the world.

What this amounts to is a *reductio ad absurdum* of the view that theories are merely descriptions of certain structures realized in the world. Indeed, the criticism just made was already aired by M.H.A. Newman (1928) as a criticism of Bertrand Russell's views. Russell could not in the end find any viable way out of the difficulty. This instructive episode has been recently recounted by van Fraassen (1997). Somewhat surprisingly, the chastened Russell nonetheless did not recognize the structuralism issue when he reviewed Ramsey's papers (see Russell 1931).

By a presumed parity of cases, it might be thought that ordinary Ramsey sentences are trivial by the same token. This criticism is not valid, however. First and foremost it is undeniable that an existential sentence like (2) has the same consequences concerning O_1, O_2, ... as a sentence like (1) that makes a statement about some particular properties and relations H_1, H_2, ... It is therefore not trivial. Admittedly, philosophers have constructed examples in which the Ramsey sentence (2) is trivial even though (it is claimed) the original theory (1) is not (cf. e.g. Achinstein 1968, pp. 84-85). A closer examination nevertheless quickly shows that in such cases the original theory (1) has no purely observational consequences to begin with, and consequently is in a sense trivial after all. Of course, it need not be trivial in another sense, viz. in that according to it theoretical and observational concepts are combined in such a way as to form a certain definite "nontrivial" structure. But this requirement need not impose any conditions on the observational part of this structure. In such cases, it should not be surprising, much less objectionable, if the Ramsey sentence is vacuous.

Speaking generally, what a Ramsey sentence says is that the structure formed by the observable concepts of the given theory can be imbedded in a structure which is a model of the entire theory. This is not normally a trivial requirement. When it is, it only means that the given theory was structurally trivial, as far as observationally trivial.

As a special case, what a super-Ramsey sentence says is that an empty set can be embedded in a structure which is a model of the given theory. Not surprisingly, this is a trivial requirement, for it depends only on the cardinality of the domain of the model.

What I have called super-Ramsey sentences are nevertheless interesting in other respects. Such a sentence imposes only minimal conditions on the world and is therefore quite weak in deductive power. It may nonetheless be extremely interesting in other respects. It says that a structure of a certain kind is realized in the world. And this structure can be extremely complicated and extremely interesting. And yet it is described without any predicates and hence apparently without any concepts whatsoever, other than the logical ones, including identity and quantifiers.

But is this really paradoxical? After all, super-Ramsey sentences contain lots of predicate variables. What they miss are only predicate constants, it may be alleged. And, as was emphasized above, such higher-order quantifiers can carry a heavy ontological load. Their use presupposes the existence of all their values which in effect include all possible properties and relations. It is not surprising that by means of such a potential wealth of concepts we can specify a lot of intricate structures. What we cannot do without higher-order constants is to locate such structures at some particular nook or corner of the actual world (or any possible world). Hence there is nothing paradoxical here. It is the second-order existential quantifiers that bring in the theoretical concepts.

4. IF FIRST-ORDER IDENTITY LOGIC

This response is totally inadequate, however. Indeed, someone might here try to turn the tables — or at least my own arguments — against me. Super-Ramsey sentences and indeed ordinary Ramsey sentences are what logicians are wont to call Σ^1_1 (sigma-one-one) sentences. That means that they consist of an initial string of second-order existential quantifiers followed by a first-order formula.

Now it is known that any Σ^1_1 sentence can be translated into an independence-friendly (IF) *first-order* sentence with the same nonlogical predicate constants. Since there are in the case of super-Ramsey sentences no nonlogical predicate constants, their first-order translations are sentences of what I will call IF first-order identity logic. Its formulas are formed by using identity symbol = as the only primitive predicate.

We will call such sentences IF first-order super-Ramsey sentences. In the case of ordinary Ramsey sentences, their translations — naturally called IF first-order Ramsey sentences — are sentences of IF first-order language with the observational concepts of the given theory as its only nonlogical primitive predicates or functions.

These statements need several explanations. First, what is independence-friendly logic? It is prompted by the insight that the ordinary first-order logic rules out certain perfectly possible patterns of dependence and independence between a number of quantifiers. (See here Hintikka 1995(b), 1996 ch. 3.) They can be restored for instance by having a notation $(\exists x/\forall y_1, y_2, ...)$ which exempts a quantifier from the scope of other quantifiers $(\forall y_1)$, $(\forall y_2)$, ... This results in a first-order logic which is much stronger than the received first-order logic but which obviously has a strong claim to being the most natural logic of quantifiers.

It can be shown that the Σ^1_1 fragment of any second-order language can be translated into the corresponding IF first-order language (see e.g. Hintikka 1996, pp. 61-63). Here correspondence means having the same primitive predicates. Since super-Ramsey sentences have only the identity as a constant predicate, they can be translated into the language of what I have called IF first-order identity logic.

This logic is highly interesting in its own right. However, it has not been studied systematically in the earlier literature. IF first-order identity logic is a *first-order* logic. What that means is that no higher-order concepts are involved in it, at least on the face of things. Such a logic does not enable us to say anything about any particular matters of fact. It is good only for purely structural descriptions. It is in a sense the logic appropriate for the study of pure structures per se devoid of all content.

In a model-theoretical perspective, what a theory does is to specify the class of its models (realizations). These can have different structures. Hence a theory also specifies a class of structures. What has been found shows that any such class of structures that can be specified by a first-order formula (IF or ordinary) can be specified by a sentence of IF first-order identity logic. This is true for instance of Peano arithmetic and of axiomatic set theory.

In short, any structure that can be specified by a first-order theory can be specified by means of IF first-order identity logic. And by means of the line of thought I have presented elsewhere, it can be argued that the same is true of any structure specifiable by means of higher-order logic. (See Hintikka 1996, ch. 9.) This includes virtually all classes of structures studied in mathematics. In other words, virtually any structure that has been considered in mathematics or science can be specified by means of a super-Ramsey sentence. The fact that we have to move to what I have called extended IF identity logic does not change the theoretical impact of our results, for the extension merely means the introduction by a fiat of the usual contradictory negation into the language in question — to the extent it can be introduced.

Since logical consequence relations depend only on the structures of the models logical formulas admit and exclude, all relations of logical consequence are manifested as logical consequence relations between formulas of IF first-order identity logic. This logic is therefore very nearly a realization of Russell's aim in *The Principles of Mathematics*,[2] where he envisaged a language containing only logical constants in which all logical and mathematical truths can be expressed.

Of course, as far as we do not try to say where in one's universe of discourse the relevant structures are located, the possibility of specifying them depends only on the cardinality of the universe. Hence IF first-order identity languages can impose only weak requirements on the world. But this is a trivial corollary of the semantical situation and it does not cast any shadows on the interest of IF identity logic. It can specify an extremely large variety of structures on the world, but it cannot so to speak tell us where in the world those structures are located. What is especially striking: any satisfiable formula, expressed in any higher-order language, has as its twin a satisfiable formula in IF first-order identity logic.

The transformation of a sentence into IF identity language does not always preserve its status vis-à-vis logical truth. We can study the logical truth of different sentences in another way, however. Instead of eliminating nonlogical constants in favor of existentially quantified variables, we can replace them by free variables, which preserves logical truth. (Of course, the result is not a sentence of IF identity logic any longer, even though it can be an IF formula.) As an example, consider the following instance of the axiom of choice

(6) $(\forall x)(\exists y)A(x,y) \supset (\exists f)(\forall z)A(z,f(z))$

where A is a primitive two-place relation. Its IF first-order translation is obtained through the following intermediate step:

(7) $(\exists f)(\exists x)(\forall y)(\forall z)(\forall u)((f(z) = u) \supset (A(x,y) \supset A(z,u)))$

The translation into an IF language is

(8) $(\forall y)(\forall z_1)(\forall z_2)(\exists x/\forall y, \forall z_1, \forall z_2)(\exists v_1/\forall y, \forall z_2)(\exists v_2/\forall y, \forall z_1)$
$(((z_1 = z_2) \supset (v_1 = v_2)) \,\&\, (A(x,y) \supset A(z_1, v_1)))$

which is easily seen to be equivalent with

(9) $(\exists f)(\exists x)(\forall y)(\forall z)(A(x,y) \supset A(z, f(z)))$

and hence with (6). This example has some independent interest. For (8) is indeed a purely logical truth in the strongest possible sense of the word, in the sense of involving only first-order logical notions. Hence (6) is a logical truth. Since the same line of thought can be applied to any instance of the axiom of choice, we have reached a powerful argument for considering the axiom as a purely logical truth. After all, the logical truth of (9) depends only on the meanings of quantifiers.

This is not just another way of looking at the axiom of choice. The same argument can be used even more generally. It can be used to vindicate the implication from any first-order sentence S_1 to the second-order sentence S_2 asserting the existence of its Skolem functions. All such conditionals

(10) $(S_1 \supset S_2)$

are logical truths.

However, they may include somewhat unfamiliar truths. For consider the set-theoretical translations S_s of the sentences S_2. By a set-theoretical translation, I mean of course a sentence where second-order quantifiers are replaced by first-order quantifiers restricted so as to range over the appropriate set-theoretical entities. For instance, the set-theoretical translation of

(11) $(\exists f)(\forall x)\, F[x,f(x)]$

will be

(12) $(\exists\alpha)$ ((α is a function) & $(\forall\beta)\, F[\beta, \alpha(\beta)]$

where "α is a function" and "$\alpha(\beta)$" are to be explicated in the usual set-theoretical way.
The generalized set-theoretical conditionals

(13) $(S_1 \supset S_s)$

are not all theorems of axiomatic set theories. One reason is that when F is complex, the classes

(14) $\{y\colon F[b,y]\}$

for different individuals b need not be sets. Alternatively, the class of all the classes (14) may not be a set. Hence the axiom of choice in all set-theoretical incarnation does not always yield the conditionals

(15) $(S_1 \supset S_s)$

as its consequences.

Thus an examination of IF first-order logic suggests considering certain new axioms of set theory which are recommended to us as set-theoretical translations of certain purely logical truths.

5. FIRST-ORDER QUANTIFIERS CAN INTRODUCE STRUCTURAL (THEORETICAL) CONCEPTS

To return to my main business, what can now be said of different kinds of Ramsey sentences? For one thing, it might now seem that a kind of Ramsey elimination is always possible. If we allow in the definition of strong Ramsey eliminability a reduction to a first-order theory that may contain IF quantifiers, we reach a notion of eliminability that I will call strong IF Ramsey eliminability.

What has been seen implies that *strong IF Ramsey eliminability*, unlike the original strong Ramsey eliminability *is always possible*. This result has several important repercussions. Among other things, the possibility of translating all Σ^1_1 sentences into the language of IF first-order logic might also seem to make nonsense of my earlier arguments against considering Ramsey reducts as effecting a genuine elimination of theoretical concepts. My suggestion was that the prima facie absent theoretical properties and relations were alive and well in (2) as values of the second-order quantifiers. But in an IF Ramsey sentence, that is, in a first-order translation of Σ^1_1-sentence (2) there are no second-order quantifiers, only first-order ones. Nor are there any other notions there that could import theoretical concepts into IF Ramsey sentences. For the IF first-order logic that facilitates such strong IF Ramsey reduction does not contain any new notions beyond ordinary first-order logic. All that happens is that our old friends, the usual first-order quantifiers, are allowed to interact more freely. The impression that theoretical concepts are not strongly Ramsey-eliminable was due only to Frege's mistake in restricting unnecessarily the way quantifiers may operate. Or so it seems. Thus the success of IF strong Ramsey elimination inevitably re-opens the entire question whether the Ramsey move is really an elimination of theoretical concepts.

Answers are forthcoming to these questions. The right way of thinking of Ramsey "elimination" is to recall the great principle of all logical analysis: there is no free lunch. One cannot get something from nothing, and conversely one cannot reduce something to nothing. What the observations that were just made really show (among other things) is that in a sense the concepts that are apparently dispensed with in Ramsey reduction are in reality alive and well in Ramsey sentences, even in their IF first-order translations. And what saying this obviously presupposes is a challenge to the received way of thinking about concepts and about the ways they are expressed in language. Usually, it is thought that substantial empirical concepts are introduced into a first-order language by its primitive predicates (including functions), *and by them only*. Logical symbols are thought of as being mere auxiliaries, facilitating the construction of meaningful sentences out of the primitive concepts. What has been found shows that this picture is wrong. What was argued above was that theoretical (structural) concepts are not introduced into theories only by theoretical constants, but also by higher-order quantifiers having such concepts (or their extensions) as values. But it can now be seen that even this observation does not go nearly far enough. Insofar as the characterization of different structures is concerned, *combinations of interacting first-order quantifiers can also introduce nontrivial theoretical concepts into a* (formal but interpreted) *language.*

The idea that quantifiers (more precisely, quantifier combinations) can introduce substantial concepts into a theory might seem to be strange. It can perhaps be made less surprising by recalling the close connection there is between quantifiers and certain kinds of functions, known as Skolem functions. These functions are, intuitively speaking, functions that tell me how to pick out suitable "witness individuals" that verify a sentence. For a simple instance, take a sentence of the form

(16) $(\forall x)(\exists y)\, S[x,y]$

What it says is that it is possible to choose, for any given value of x, a value of y such that S[x,y]. If we think of this choice as being implemented by a function f, we see that (6) is equivalent with

(17) $(\exists f)(\forall x)\ S[x,f(x)]$

What this means is that the dependent quantifier $(\exists y)$ in (16) introduces in effect the choice function f.

Thus concepts are imported into our discourse not only by higher-order constants and by higher-order quantifiers, but also by dependent *first-order* quantifiers.

These results have a major impact on discussions of the contrast between observation terms and theoretical terms (and of their interplay with each other in a scientific theory). These discussions have been conducted invariably in terms of a dichotomy between two kinds of primitive concepts that into a theory or, equivalently, in terms of two kinds of primitive nonlogical predicates of the theory in question. Now it is seen that this picture has to be modified. It is not even sufficient to consider higher-order quantifiers and their values. The contributions of (patterns of) first-order quantifiers have also to be considered.

For instance, we can now reach a perspective on the failure of the original strong Ramsey eliminability. Assume that T=(1) is the original theory and T* its strong IF Ramsey reduct. Why does it not also have an ordinary strong Ramsey reduct T**? The answer is that IF first-order logic is stronger than ordinary first-order logic that must be used in T**. But this is merely a matter of the strength of the logics used in T* and in T**. Hence the failure of (ordinary) strong Ramsey eliminability does not tell anything about how theoretical concepts unrelated to quantifiers fare in first-order theories.

Indeed, the examples in the literature of theories that do not have strong (ordinary) Ramsey reducts can all be transformed into examples of IF first-order formulas that do not have any ordinary first-order equivalents. This is the case for instance of the example (due to Dana Scott) presented by Sneed (1971, p. 54). It turns on the question whether there is a one-to-one mapping between two classes of individuals. The existence of such one-to-one mappings cannot usually be expressed by means of ordinary first-order logic, but it can easily be expressed by means of the resources of IF first-order logic.

Of course, the only kind of force that quantificationally introduced concepts have is purely structural. But then so is (as Ramsey realized) the force of theoretical concepts in the first place.

A more general consequence of these observations is that the entire discussion about theoretical concepts in scientific theories and about their possible elimination must be reconsidered. For the possibility that theoretical concepts might be introduced by patterns of first-order quantifiers has not been taken into account for this discussion. For one thing, the results reached here should put to rest for good the idea that Ramsey sentences offer us an elimination of theoretical concepts in any natural sense of the term.

The fact that theoretical concepts have not really been eliminated from IF first-order Ramsey sentences is illustrated by the fact that they can be so strong that they are not equivalent with any ordinary first-order sentence.

All this can be illustrated further by a simple example. Let

(18) $(\forall x)(\forall z)\, S[x,f(x),z,g(z)]$

be a (first-order) theory. Its Ramsey sentence is

(19) $(\exists f)(\exists g)(\forall x)(\forall z)\, S[x,f(x),z,g(z)]$

The IF first-order translation of (2) is

(20) $(\forall x)(\forall z)(\exists y/\forall z)(\exists u/\forall x)\, S[x,y,z,u]$

There are examples where S is such that (20) does not have any ordinary (unslashed) first-order equivalent.

Thus an apparent elimination of f and g from (18) is possible. However, this elimination is in a deeper perspective illusory. For the choice functions f and g survive in (20) in the form of quantifier patterns. For instance, notice how the pattern of dependencies and independencies in (20) determines what arguments the functions f and g have, in this case, that the former depends only on x and the latter only on z.

These results throw interesting light on the nature of quantifiers. The way in which first-order quantifiers introduce concepts into our discourse is by as it were deputizing their Skolem functions. Indeed, these Skolem functions *are* the concepts they introduce. But Skolem functions do not play any role in the most popular theories of quantifiers. If quantifiers are construed as higher-order predicates or if they are thought to receive their meaning by "ranging over" a range of values or to receive it from the totality of their substitution-values, the kind of interplay of different quantifiers with each other that is codified by their associated Skolem functions remains unexplained. Hence the discovery of the genuine introduction of new concepts by first-order quantifiers in the form of their Skolem functions provides strong reasons for such theories of quantifiers as can do justice of their function of codifying certain choice functions (Skolem functions). Such theories include Hilbert's treatment of quantifiers by means of his epsilon-symbol and my game-theoretical semantics.

Further reasons to the same effect are provided by the vindication of the purely logical status of the axiom of choice via a translation to IF first-order identity logic. For this logic is merely a way of spelling out the consequences of the meaning of quantifiers and identity.

In conclusion, Ramsey sentences do not implement any genuine elimination of theoretical concepts. Appearances notwithstanding, the same goes for their IF first-order translations. What is remarkable about them is that they show how such concepts are introduced into our discourse by quantifiers and thereby what the real meaning of quantifiers is.

ENDNOTES

[1] On the other hand, for the validity of the step from (1) to (2) we do not have to make any other assumptions concerning the domain of individuals. Hence it is not correct to say that "Ramsification swells the ontology of the original theory" (Tuomela 1973, p. 66). In any case, the domain of interpretation of (1) and (2) has to be the same, since they have to be interpreted by reference to a given domain.

[2] Later, abandoned; see Russell (1931).

REFERENCES

I am referring to Ramsey's papers both in the 1931 and in the 1978 edition. They are:

Ramsey, F.P., 1931, *The Foundations of Mathematics and other Logical Essays*, edited by R.B. Braithwaite, Routledge and Kegan Paul, London.

Ramsey, F.P., 1978, *Foundations: Essays in Philosophy, Logic, Mathematics and Economics*, edited by D.H. Mellor, Routledge and Kegan Paul, London.

Achinstein, P., 1968, *Concepts of Science*, the Johns Hopkins Press, Baltimore.

Carnap, R., 1966, *Philosophical Foundations of Physics*, Basic Books, New York.

Hintikka, J., 1995(a), "Standard vs. Nonstandard Distinction: A Watershed in the Foundations of Mathematics," in Jaakko Hintikka, editor, *From Dedekind to Gödel: Essays on the Development of the Foundations of Mathematics*, Kluwer Academic, Dordrecht, pp. 21-44.

Hintikka, J., 1995(b), "What Is Elementary Logic?," in *Physics, Philosophy and the Scientific Community*, K. Gavroglu et al., editors, Kluwer Academic, Dordrecht, pp. 301-326.

Hintikka, J., 1996, *The Principles of Mathematics Revisited*, Cambridge University Press.

Moore, G.H., 1988, "The Emergence of First-Order Logic," in *History and Philosophy of Modern Mathematics*, William Aspray and Philip Kitcher, editors, University of Minnesota Press, Minneapolis, pp. 95-135.

Newman, M.H.A., 1928, "Mr. Russell's Causal Theory of Perception," *Mind* vol. 37, pp. 137-148.

Papineau, D., 1996, "Theory-dependent Terms," *Philosophy of Science* vol. 63, pp. 1-20.

Russell, B., 1931, "Critical Notice of F.P. Ramsey, *The Foundations of Mathematics and Other Essays*," *Mind* vol. 40, pp. 476-482.

Sahlin, N.-E., 1990, *The Philosophy of F.P. Ramsey*, Cambridge University Press.

Sneed, J.D., 1971, *The Logical Structure of Mathematical Physics*, D. Reidel, Dordrecht.

Stegmüller, W., 1970, *Theorie und Erfahrung (Probleme und Resultate der Wissenschaftstheorie und Analytischen Philosophie, Band II)*, Springer-Verlag, Berlin. (See chapter 7.)

Tuomela, R., 1973, *Theoretical Concepts*, Springer-Verlag, Wien.

van Benthem, J., 1978, "Ramsey Eliminability," *Studia Logica* vol. 37, pp. 321-336.

van Fraassen, B., 1997, "Structure and Perspective: Philosophical Perplexity and Paradox," in *Logic and Scientific Methods*, M.L. Dalla Chiara et al., editors, Kluwer Academic, Dordrecht, pp. 511-530.

14

TOWARDS A GENERAL THEORY OF IDENTIFIABILITY

1. WHAT IS IDENTIFIABILITY?

What is identifiability, anyway, and what does it have to do with definitions and definability? The basic intuitive idea is clear. A concept (say, a one-place predicate P) occurring in a theory $T[P]$ is *definable* on the basis of this theory iff the theory determines the interpretation of P as soon as the interpretations of the other concepts occurring in $T[P]$ are fixed. More explicit, this is what, the definability of P on the basis of $T[P]$ means.

What is meant by determination here remains to be spelled out. Furthermore, there are of course other variants of the idea of definability, some of which will be characterized below.[1] These questions nevertheless need not detain us here.

It may happen, however, that the interpretation of P is not determined on the basis of the theory alone, but is determined by the theory *together with* a number of auxiliary *empirical results*, such as observations or experimental outcomes.

Then it is said that P is *identifiable* on the basis of $T[P]$. Such a notion of identifiability is unlike the notion of definability in that it is relative to a model M (of the language of $T[P]$), for the auxiliary empirical information has to pertain to some one "world", i.e., some model of the language of $T[P]$. (When I mention "models" in the rest of this paper, I shall mean models of the language of the underlying theory, unless something else is indicated.) We can make the notion of identifiability independent of M by so to speak integrating over the space of models, e.g., by stipulating that P is identifiable on the basis of $T[P]$ *simpliciter* iff it is identifiable in each model of $T[P]$.

The notion of identifiability is also relative to the class of possible observations, as it were to the totality of answers that nature can give to an inquirer's questions.

Throughout this paper, I shall use the definability or identifiability of a simple one-place predicate as an example. All the definitions and results are easily extended to other cases.

There are distinctions one has to make between different variants of identifiability. Some of them have parallels in the theory of definability as distinctions between different kinds of definability.

2. WHY DOES IDENTIFIABILITY MATTER?

Why is the notion of identifiability important? Several weighty reasons can readily be found.

First, the notion of identifiability has played a significant role in the actual methodology of certain empirical sciences. Indeed, it first caught theoreticians' eye in the context of actual theorizing in mathematical economics.[2] It is not hard to guess in the light of hindsight what the problem was. Suppose a concept (a parameter) like our predicate P occurs in a theory $T[P]$ which is intended for the actual mastery of empirical phenomena. For such a purpose, the theory $T[P]$ is useless unless the value of the parameter in question can be estabilished by means of auxiliary exmpirical information (observable data). Yet if the logical (mathematical) structure of $T[P]$ is of the right sort (or, from the vantage point of the intended use of the theory, of the wrong sort), this may be intrinsically impossible (impossible for logical and mathematical reasons). This is precisely what is meant by the nonidentifiability of the parameter in question. And this is precisely what happened with certain macro-economic theories. In order to be forewarned (and forearmed), econometricians have since developed methods of recognizing the conditions on which such nonidentifiability can or cannot occur, at least in the types of theories most frequently used in economics.

A simple example of an identifiability problem comes up when the underlying theory assumes the form of a pair of linear equations

$$(1) \qquad \begin{aligned} a_1x + b_1y &= 1 \\ a_2x + b_2y &= 1 \end{aligned}$$

if the determinant D

$$D = \begin{vmatrix} a_1 & b_1 \\ a_2 & b_2 \end{vmatrix}$$

is different from zero, the only observable values of the variables x and y are the solutions to the pair of equations (1). But, from such observations one can never determine the parameters a_1, b_1, a_2, b_2.

However, of $D = 0$ the two equations (1) do not have any solutions unless they are equivalent. Then the parameters a_1 and b_1 can be identified from observations concerning two pairs of values of the variables x_1, y_1, and x_2, y_2, i.e., as solutions of the following equations with respect to a_1 and b_1:

$$(2) \qquad \begin{aligned} a_1 x_1 + b_1 y_1 &= 1 \\ a_1 x_2 + b_1 y_2 &= 1 \end{aligned}$$

In spite of its simplicity, this example is characteristic of the conceptual structure of identification problems in econometrics.

I shall return below to the role of the concept of identifiability in other sciences. This concept belongs in any case squarely to the actual working methodology of several different actual empirical sciences.

3. IDENTIFIABILITY DISGUISED AS DEFINABILITY

Second, what is called in philosophical or methodological discussions "definability" frequently, perhaps typically, turns out to be identifiability and not definability in any of the several senses of definability studied by logicians. Hence the study of identifiability is an integral part of any comprehensive study of definability as the latter term is actually used by philosophers and methodologists.

In some cases, the assimilation of definability to identifiability has been made in so many words. For instance, Herbert Simon acknowledges the fact that he uses the term "definability" to all practical purposes as a synonym for (a generalization of econometricians' concept of) identifiability.[3]

There is nothing intrinsically wrong with such a usage. We shall in fact see that there are close ties between the notions of identifiability and definability. (*Vide* theorems 11 below.) However, there is little payoff in the employment of the notion of identifiability on a general philosophical or methodological level as long as there does not exist a general logical theory of identifiability. Such a theory was not available at the time Simon was writing. In a sense, my paper is one of the first serious attempts to develop a general theory of identifiability.

A comparison with definability is instructive here. Even though the majority of philosophers seem to be unaware of it, there exists a rich logical theory of definability which has a great deal of potential philosophical interest. This theory comprises among other things definitions of different types of definability, their interrelations, their criteria, etc. In contrast, there does not exist any equally explicit general theory of identifiability. Admittedly, it turns out that certain aspects of the general theory of definability are in fact highly relevant to the study of identifiability, too. This fact has its deep reasons in the logical structure of identifiability problems. But even these existing results have not been collected together and presented with anything like their proper motivation. One of the purposes of the present paper is to show a general logical theory of identifiability can be developed as an extension and as a partial analogue to the general logical theory of definability.

To return to the relation of identifiability and definability, another danger in using "definability" simply as another term for identifiability with unnecessary disagreements as to what definability "really is" or how scientists "actually use" the notion of definability. This risk is not merely an abstract possibility, either. The one-time protracted controversy between Herbert Simon and Patrick Suppes concerning the role of definitions in science turns out, in the light of 20/20 hindsight, to have been a classic case of a debate at cross-purposes.[4] Suppes was using the terminology of definitions and definability in its established logical sense, whereas Simon employed the same terminology to mean some sort of generalized notion of identifiability for which there did not exist at the time any exact theory (outside the special cases studied by the likes of econometricians).

Simon was undoubtedly right in one important respect, however. When the terms "defining", "definition", and "definability" were used in traditional foundational or methodological discussions, what was meant was usually identifiability, not definability. For instance, when Ernst Mach posed the problem as to whether the concept of mass is definable in Newtonian mechanics, he was not asking whether the fundamental equations of (a suitable formulation) Newtonian mechanics logically imply an explicit definition of mass in terms of the other priorities of Newtonian mechanics, nor was he inquiring into any other variety of definability of the concept of mass on the basis of the fundamental laws of classical mechanics.[5] What Mach was asking was whether these laws enable a physicist to determine in each and every case the value of a point mass by means of suitable observations. In other words, Mach's question concerned, not definability, but identifiability.

4. THE INTERROGATIVE MODEL OF INQUIRY

How can we develop an explicit theory of identifiability? As was indicated above, it is technically possible to discuss some problems of identifiability simply by using in a suitable way the logical techniques employed in the existing logical theory of the different kinds of definability. This procedure would make it very hard to see the rationale of the different conceptualizations that have to be used here, and it would not be helpful heuristically, either, in bring out partial analogies between identifiability and definability. A more natural framework is offered by the *interrogative model of inquiry* which I have developed.[6] It is calculated to be a general conceptual tool in different kinds of investigations into epistemology and philosophy of science.

The interrogative model can be formulated in the form of a game. This game is what game theorists could call a game against nature, in that it is a two-person game in which only one player ("the Inquirer") is active whereas the other ("Nature") comes in only (or mainly) as a source of answers to the Inquirer's questions. In this paper, it will be assumed that all the answers are true (in the model to which the inquiry pertains).

At each stage of the game, the Inquirer has a choice between two kinds of moves, (i) a logical move (also called an inference move or *tableau*-building move) and (ii) an interrogative move. The former consists essentially in the Inquirer's drawing a deductive inference from earlier results. The latter consists essentially in the Inquirer's putting a question to Nature. If Nature comes up with an answer, this answer can be used by the Inquirer as an additional premise.

In order to be able to formulate the rules form these moves more explicitly, we can decide to use as our bookkeeping method — or, perhaps rather, the Inquirer's bookkeeping method — a variant of E. W. Beth's method of semantical *tableaux*.[7] Then the logical moves are steps of *tableau* construction. In speaking of them, we can use all the usual vocabulary pertaining to Beth's *tableaux*, such as the left column, the right column, *subtableau*, closure, etc. Since the basic idea of the *tableau* technique is extremely simple, using this technique will enable the reader to appreciate the interrogative model even without much knowledge of logic. Very briefly, a *tableau* which begins with T in the left column and C in the right column is the record of a thought-experiment, viz. an attempt to think of T as true and C as false in the same "world" (model). If such an attempt is frustrated in all directions (i.e., if the *tableau* closes), C cannot fail to be true if T is, i.e., C follows logically from T.

Using the *tableau* technique of course means assuming that we are dealing with (applied) first-order languages. I shall make this assumption for the purposes of this paper. Since we are dealing with interpreted languages, we have to make a sharp distinction between real ("proper") names (of individuals in the domain $do(M)$ of the model M to which the inquiry pertains) and "dummy names" introduced by the logical instantiation rules.

An interrogative game begins with a *tableau* where the only entry in the left column is an initial premise T. (It can of course also be a set of premises.) The only entry in the right column initially is the conclusion C to be proved. (It can also be a set of conclusions whose disjunction is to be proved.)[8]

In this paper, it is normally assumed that T is true in the model in question.

5. RULES FOR LOGICAL MOVES

I shall nevertheless modify even the rules for logical moves somewhat. I shall first give the rules and then indicate the rationale for the modifications.

The *tableau*-forming rules can be summarized as follows:

What occurs in the left column of a *subtableau*.	What may be added to the left column of the same *subtableau*.
(L.&) $(S_1 \,\&\, S_2)$	S_1 and/or S_2
(L.E.) $(\exists x)\, S[x]$	$S[a]$ ("a" = a new dummy name)
(L.A.) $(\forall x)\, S[x]$	$S[b]$ (for any *dummy name* or *real name* "b" occurring in the *subtableau*); $S[y]$ for any variable y which occurs free in the subtableau.

(L. ~ ∨) ~ $(S_1 \vee S_2)$	~ S_1 and/or ~ S_2
(L. ~ E) ~ $(\exists x)\, S[x]$	~ $S[b]$ (b as in (L.A.))
(L. ~ A) ~ $(\forall x)\, S[x]$	~ $S[a]$ (a as in (L.E.))
(L. ∨) $(S_1 \vee S_2)$	S_1 or S_2 (split the construction into two further *subtableaux*, with S_1 and S_2 as the new members).
(L. ~ &) ~ $(S_1 \,\&\, S_2)$	~ S_1 or ~ S_2 (as in L. ∨))

(R.&), (R. ∨), (R.A), R. ~ &), (R. ~ ∨) and (R. ~ E) are simply duals of the correspond rules on the left.

Occurring in the right column of a *subtableau*.	Add to the right column of the *subtableau*.
(R.E.) $(\exists x)\, S[x]$	$S[b]$ (for any dummy or real name b occurring in the same *subtableau*); $S[y]$ for any variable y which occurs free in the *subtableau*.
(R. ~ A) ~ $(\forall x)\, S[x]$	~ $S[b]$ (b as in (R.E))

Closure rules:

(C.T) A *tableau* is closed iff all its *subtableaux* are closed.

(C.L) A *subtableau* is closed if S and ~ S occur in its left column (for any S).

(C.R) A *subtableau* is closed if S and ~ S occur in its right column (for any S).

(C.LR) A *subtableau* is closed if S occur in both of its columns (for any S).

By way of an explanation of these rules, and especially of their difference from the usual *tableau* rules, the following can be said:

The basic idea of the *tableau* method is to see if the purported conclusion C actually follows from T by assuming that T is true but C is false and then by reducing this counter-assumption *ad absurdum*. Hence everything in a left column is thought of as being true while everything in a right column is thought of as being false. The usual *tableau* rules can be based on this idea. But in an interrogative inquiry something else is also assumed. It is assumed that what is in a left column is *known* to be true. For this reason nothing can be moved from the right column to the left one. For this reason, no traffic from the right to the left can be allowed. Hence each column must unfold by itself.

6. RULES FOR INTERROGATIVE MOVES

Rules for question-answer moves make use of the notion of the *presupposition* of a question. This notion does not have to be analyzed here, however. In interrogative moves, all action takes place in the left column.

(Q.P) If a disjuction $(S_1 \vee S_2)$ occurs in the left column of a *(sub)tableau*, the Inquirer may use it as a presupposition of the propositional question

$$(S_1 \text{ or } S_2)?$$

The two possible answers are S_1 and S_2. If Nature gives one of the answers, it is added to the left column of the same *(sub)tableau.*

(Q.Wh) If $(\exists x)S[x]$ occurs in the left column of a *(sub)tableau*, the Inquirer may use it as the presupposition of the wh-question.

Who (what, where, when, . . .), say x, is such that $S[x]$?

or in general,

Which individual, call it x, is such that $S[x]$?

The possible answers are of the form

$$S[b]$$

where "b" is a name of a member b of the domain $do(M)$ of the model

M to which the inquiry pertains. If Nature gives an answer of this form, it is added to the left column of the same *(sub)tableau*.

Because the table construction may branch, it may happen that the presupposition of a legitimate question is false. Then no true answer is available. In such cases, Nature is allowed to give any possible answer (of the right form). The same holds if the same column contains an earlier entry which is not true. Otherwise, Nature's answer must be true.

Since dummy names do not have any specific bearer, none of Nature's answers is allowed to contain dummy names. The same goes for free variables.

7. GENERAL REMARKS ON INTERROGATIVE DERIVABILITY

If the Inquirer has managed to close the game *tableau*, it is said that she or he *has derived C from T interrogatively*. If the Inquirer has a strategy in the interrogative game by means of which she or he can win (close the *tableau*) no matter what Nature does, it is said that *C is interrogatively derivable from T*. In other words, C is interrogatively derivable from T is and only if the Inquirer has a strategy which wins against any strategy of Nature's.

It is in order to point out that this does not mean that the *tableau* construction (which includes all of Nature's admissible answers) comes to an end after a finite number of steps. Even if each branch closes, there may nonetheless be arbitrarily long branches, because Nature can sometimes choose her answers from an infinite set of possible answers.

The need of considering several *subtableaux* can be eliminated by replacing the branching rules (L. $\vee$) and (R.&) by the following:

(L. $\vee$ .Nature) When $(S_1 \vee S_2)$ occurs in the left column, the Inquirer may ask Nature to add S_1 or S_2 to the left column.

(R.&.Nature) When $(S_1 \,\&\, S_2)$ occurs in the right column, the Inquirer may ask Nature to add S_1 or S_2 to the right column.

In both rules, Nature must comply. The choice of the formula to be added is up to Nature in both cases. It is not hard to see that this change does not alter the notion of interrogative derivability. The modified version turns out to be heuristically useful in many proofs.

That C can be derived interrogatively from T in M (by reference to a fixed set of restrictions on Nature's answers) will be expressed by

(1) $M: T \vdash C$

From the definitions of the interrogative games above it can be seen in what sense (1) is a generalization of the usual relation of logical consequence

(2) $T \vdash \mathrm{C}$

In an extreme case, the Inquirer asks no questions (and/or Nature answers none). Then (1) and (2) coincide. Otherwise, (2) entails (1) but not always *vice versa*.

The notion of interrogative derivability has considerable philosophical interest. It is fairly obvious that what is meant by "consequences" and even by "deductive consequences" in contemporary philosophy of science is better captured by interrogative derivability. For instance, in expounding the so-called hypothetico-deductive model of science philosophers do not always realize how little is actually derivable purely deductively from an abstract scientific theory. I suspect that what the proponents of the model have intended is really a hypothetico-interrogative model of a science rather than a hypothetico-deductive model.

The interrogative derivability of C in a model M of T is in an interesting way intermediate between *a logical implication* from T to C and the *truth of C in M*. If questions play no role, we have the former. If questioning is pursued under optimal conditions (in a certain specifiable sense), C is interrogatively derivable in M without any T iff C is true in M.

One of the ways in which interrogative derivability differs from deductive consequence (logical derivability) is that the interrogative counterparts to cut elimination and similar proof-theoretical results do not hold for it.[9] Accordingly, the following game rule will add to the purview of interrogative derivability:

(L.Taut.) At any stage of the game, the Inquirer may add any tautological disjunction

$$(S \vee \sim S)$$

to the left column.

The dual rule will be called (R. Cont.). It entitles the Inquirer to add any contradictory conjunction (S & $\sim S$) to the right column.

The more complex the sentence S is quantificationally (e.g., the more layers of unlike quantifiers it contains), the more extra power is yielded by an application of (L.Taut.).

As I have shown elsewhere, the force of an application of (L.Taut.) is to assume that the Inquirer is prepared to raise the question whether or not it is the case that S.[10] Hence applications of (L.Taut.) by the Inquirer amount to extending the range of the Inquirer's awareness.

This observation suggests that we can safely assume a number of special cases of (L.Taut.). If the Inquirer has already formulated S explicitly, he or she can scarcely be unaware of the question of its truth. This motivates assuming, as I shall do in the rest of this paper, that (L.Taut.) holds if the following condition is satisfied:

> S is a subsentence of a formula in the *subtableau* in question or a substitution-instance of a subformula of a formula in the *subtableau* with respect to names occurring in the *subtableau*.

The general theory of interrogative derivability can be developed as an extension of the usual proof theory (theory of deductive derivability) and also to some extent as an extension of the usual first-order model theory. Now one of the systematic mainstays of the metatheory of the usual (deductive first-order) logic is the Interpolation Theorem (IT) due to William Craig:[11]

INTERPOLATION THEOREM (IT). If $T \vdash C$ but neither $\vdash \sim T$ not $\vdash C$, then there is an interpolation formula I such that

(a) $T \vdash I$

(b) $I \vdash C$

(c) Each free nonlogical symbol of I occurs both in T and C.

I shall not reproduce a proof of IT here. It can be proved easily by means of induction on the length of the proof of $T \vdash C$.

The Interpolation Theorem holds also when T and C are (possibly infinite) sets of formulas.

8. THE FUNDAMENTALS OF DEFINABILITY

In order to see how the usual theory of definability can be extended to identifiability, we first have to review some of the most basic concepts and results concerning definability. I shall use the definability of a one-place predicate P as the basis of a first-order theory $T[P]$ as a test case. For simplicity, it will be assumed throughout this paper that $T[P]$ has only infinite models.

EXPLICIT DEFINABILITY

$$T[P] \vdash (\forall x)(Px \leftrightarrow D[x])$$

where (i) P does not occur in the *definiens* $D[x]$; (ii) x is the only individual variable in $D[x]$; (iii) $D[x]$ contains only nonlogical constants (concepts) occurring in $T[P]$.

PIECEWISE DEFINABILITY

$$T[P] \vdash \vee_{i=l}^{i=k} (\forall x)(Px \leftrightarrow D_i[x])$$

with the same restriction on each $D_i[x]$ as on $D[x]$ in explicit definability.

In many ways, piecewise definability is the normal kind of definability (on the basis of a theory). Explicit definability can be thought of as a special case, i.e., a "uniform piecewise definability". In a complete theory, explicit and piecewise definability coincide.

IMPLICIT DEFINABILITY

(a) *Weak version*: Assume that $T[P]$ and a model M of $T[P]$ are given. Then there is for each $b \in do(M)$ (= the domain of individuals of M) a formula $F[x]$ without P such that the following are true in M:

(i) $F[b]$

Either

(ii) $T \vdash (\forall x)(F[x] \supset Px)$

or

(ii) $T \vdash (\forall x)(F[x] \supset \sim Px)$

(b) *Strong version*:

$$(T[P] \,\&\, T[P']) \vdash (\forall x)(Px \leftrightarrow P'x)$$

where $T[P']$ is like $T[P]$ except for containing a new one-place predicate P' wherever $T[P]$ contained P.

Note: This use of the term "implicit definition" seems to go back to Beth. It differs from the most usual meaning of "implicitly defined" among philosophers, which assumes that a concept is "defined" already by the role it plays in a theory like our $T[P]$.

DEFINABILITY IN A MODEL. P is definable (on the basis of the other concepts of $T[P]$) in a model M of the theory $T[P]$ if and only if an explicit definition of P is true in M.

The main results concerning these metatheoretical concepts are the following:

(1) Assume that P is weakly implicitly definable in M on the basis of $T[P]$. Then it is definable in M if and only if there is a finite upper bound to the quantificational depth (maximal length of sequences of nested quantifiers) of the different formulas F in the definition of weak implicit definability.

(2) Strong implicit definability and explicit definability coincide (Beth's Theorem).[12]

(3) Definability of P in every model of $T[P]$ is equivalent with piecewise definability (Sevenonius).[13]

The most crucial of these results is (2). It can be proved easily with the help of the Interpolation Theorem.

Notice that Beth's Theorem is very handy when one has to try to find out whether a concept is definable on the basis of a theory. The usual techniques of deduction cannot be applied directly for the purpose of testing the applicability of the definition of explicit definability, for one does not normally know ahead of time what the definiens $D[x]$ is supposed to be. In contrast, in deciding questions of implicit definability one only needs to try to deduce $(\forall x)\ (Px \leftrightarrow P'x)$ from $(T[P]\ \&\ T[P'])$. A suitable proof of the IT shows how to construct the actual definition $D[x]$ from a successful deduction.

Moreover, if one does it by means of the *tableau* technique, one can use it so that the end product is either a proof of definability or else (at least in the limit) a model in which all the other concepts of $T[P]$ and of $T[P']$ have a definite interpretation but P and P' are interpreted differently. Proving nondefinability (sometimes also known as independence) by means of such a model is known as Padua's method.[14]

Beth's Theorem shows that Padua's method is in a sense the universal method of proving nondefinability in first-order theories, and that a Padua-type model can always be constructed by means of an attempt to establish (weak) definability.

9. EXTENDING THE INTERPOLATION THEOREM

The usual (deductive) Interpolation Theorem can be extended so as to become a tool in the study of identifiability. The extended theorem can be formulated as follows:[15]

EXTENDED INTERPOLATION THEOREM. Assume

(i) $M: T \vdash C$

(ii) Not $T \vdash C$

(iii) Not $M: T \vdash (S \,\&\, {\sim} S)$ (for any S)

Then there is a formula I (the interpolation formula) and individuals $b_1, b_2, \ldots, b_k \in do(M)$ such that

(a) $M: T \vdash I$

(b) $I \vdash C$

(c) Each predicate, dummy name, and free variable in I occurs in both T and C.

(d) Each (proper) name in I *either* occurs in both T and C *or* was introduced in the proof of (i) for the first time by an answer to a wh-question *or* was imported in the same proof from the right column to the left column by an application of (L.A) or (L. ~ E)

(e) No wh-questions are used in the proof of (a).

(f) All the answers used in the proof of (a) are used in the proof of (i).

An examination of the Extended Interpolation Theorem and its proof shows how the structure of the interpolation formula partially reflects the structure of the original proof of (i). The clause (d) above shows how the nonshared names of I reflect the applications of (Q.wh) and such applications of (L.A) as import a name to the left column from the right column. Furthermore, it can seen that the number of layers of quantifiers in I equals the number of applications of (L.A) or (L. ~ E) to dummy names or free variables occurring only in the right column plus the number of applications of (R.E) (or (R. ~ A)) to real or dummy names occurring only in the left column.

One remarkable feature of the Extended Interpolation Theorem is that it holds largely independently of what restrictions are imposed on Nature's answers, as long as they remain the same throughout the theorem. Admittedly, a few minor assumptions have to be made in the Extended Interpolation Theorem as to what questions will be answered by Nature. These assumptions are extremely natural, however, and will not be examined here in any detail.

10. DEFINING IDENTIFIABILITY

Now we have finally reached a point at which we can define different kinds of identifiability. The fundamental idea is clear on the basis of the preliminary explanations offered in sec. 1 above. In the same way as interrogative derivability is a generalization of deductive derivability, identifiability is a generalization of definability. Because of this analogy, the interrogative model of inquiry suggests a number of definitions which are roughly parallel with the corresponding definitions of different kinds of definability.

EXPLICIT IDENTIFIABILITY. P is explicitly identifiable in M on the basis of *T[P]* iff

(*) $$M: T[P] \vdash (\forall x)(Px \leftrightarrow D[x, a_1, a_2, \ldots, a_k])$$

where $a_1, a_2, \ldots, a_k$ **ε** $do(M)$ and where the *definiens* satisfies the same conditions as in explicit definability (except for the presence of a_1, a_2 $\ldots, a_k$).

Moreover, (*) must be established without the help of any answers which contain P.

PIECEWISE EXPLICIT IDENTIFIABILITY

$$M: T[P] \vdash \vee_i (\forall x)(Px) \leftrightarrow D_i[x, a_1, a_2, \ldots, a_k])$$

with the help of any answers which contain P. Each D_i must of course be like D in explicit identifiability.

STRONG IMPLICIT IDENTIFIABILITY

$$M: (T[P] \,\&\, T[P']) \vdash (\forall x)(Px \leftrightarrow P'x)$$

without the help of any answers which contain P or P'

Here $T[P]$ and $T[P']$ are as in the definition of strong implicit definability.

WEAK IMPLICIT IDENTIFIABILITY. For each element b of $do(M)$, either

$$M\colon T[P] \vdash Pb \qquad \text{or}$$
$$M\colon T[P] \vdash \sim Pb$$

without the help of answers containing P.

11. BASIC RESULTS

The basic results in the theory of identifiability include the following:

THEOREM 11.1 (Extended Beth's Theorem): Strong implicit identifiability implies explicit identifiability.

This can be proved by means of the Extended Interpolation Theorem essentially in the same way as its deductive (logical) counterpart.

Many of the same things as were said above of the role of Beth's Theorem in the theory of definability can be said of the Extended Beth's Theorem in the theory of identifiability. It opens up a way of testing by means of an interrogative argument whether a concept is (explicitly or strongly implicitly) identifiable in a model on the basis of a theory. (Cf. also sec. 14 below.)

Identifiability and definability differ from each other in that weak implicit identifiability does not imply strong implicit identifiability. There nevertheless are results relating the two.

THEOREM 11.2. Explicit identifiability implies strong implicit identifiability.

THEOREM 11.3. If P is explicitly or weakly implicitly identifiable in every model of $T[P]$, then it is explicitly definable on the basis of $T[P]$.

This theorem shows why it is so easy to confuse identifiability and definability with each other.

THEOREM 11.4. If P is piecewise explicitly identifiable in each model of $T[P]$, then

$$T[P] \vdash (\exists y_1)(\exists y_2)\ldots(\exists y_m) \vee_i (\forall x)(Px \leftrightarrow D[x, y_1, y_2, \ldots, y_m])$$

This theorem is closely related to a well-known result in the theory of definability.[16]

THEOREM 11.5. Assume that $M\colon T[P] \vdash Pb$ without answers containing P. Then there are $F[x, y_1, y_2, \ldots, y_k]$, and $a_1, a_2, \ldots, a_k \in do(M)$ such that

(i) $M\colon T[P] \vdash F[b, a_1, a_2, \ldots, a_k]$ (without answers containing P)

(ii) $T[P] \vdash (\forall x)(\forall y_1)(\forall y_2)\ldots(\forall y_k)(F[x_1, y_1, y_2, \ldots, y_k] \supset P_x)$.

(iii) There are no individual constants in $F[x, y_1, y_2, \ldots, y_k]$.

(iv) P does not occur in F.

Furthermore, the same holds for $\sim Pb$ instead of Pb.

Theorem 11.5 might be compared with Hempel's covering law model of explanation. It shows that there might be a little bit more to be said for the covering law model from a logical point of view than philosophers have recently acknowledged.

THEOREM 11.6. Assume that P is weakly implicitly identifiable in M on the basis of $T[P]$. Then it is strongly implicitly identifiabile if and only if there is a finite upper bound both to k and to the quantificational depth of F_2 in theorem 11.4.

These results will, I hope, give the reader the sense of what can be done in the general logical theory of identifiability.

There is a connection between the results stated here, especially the Extended Beth's Theorem, and the study of identifiability in econometrics. Assume that $T[P]$ is a set of equations. What the Extended Beth's Theorem says then is that P is identifiable on the basis of $T[P]$ iff this set of equations is solvable for P with respect to all the different available values of the other concepts of $T[P]$. Much of the econometric theory of identifiability can be thought of as spelling out the consequences of this observation.

12. IDENTIFIABILITY, MEASURABILITY AND OBSERVABILITY

An analysis of the concept of identifiability in terms of the interrogative model can be useful in philosophy, especially in the philosophy of science.

As a small example I can mention a frequently employed argument for the alleged incommensurability of classical and relativistic mechanics.[17] This argument for the alleged incompatibility of the two theories is based on the claim, which can be granted here, that certain concepts are identified in different ways in Newtonian and Einsteinian physics. That this kind of argument carries little weight is suggested by theorems 11.5 and 11.6. The former shows that the identification of P can take place in different ways for different individuals. Theorem 11.5 says, essentially, that the several identifying formulas can be combined in one.

A larger-scale application deals with the notions of observability and measurability. Philosophers of science have discussed extensively the contrast between observable concepts and theoretical concepts, but this discussion has not produced much genuine enlightenment.[18] The interrogative model shows that more distinctions have to be made here than are customarily made in the literature. The most important distinction that has to be made is between, on the one hand, what can be said of the answerability of questions containing a given concept and, on the other hand, the identifiability of a concept, in the different senses of identifiability.

The set A of available answers normally depends on the concepts these answers contain, but of course it normally depends on other factors as well, such as the quantificational depth of the answers.

It seems to me that the terms "observable" and "observability" are most naturally taken to pertain to the availability to answers to the Inquirer's questions to Nature (including one's perceptual environment). After all, observations can be thought of as a subclass of answers put to the observer's environment. Of course, finer distinctions between different variants of observability can be made here. In one sense of observability (observability$_2$), a concept is observable iff it occurs in an available answer to the Inquirer's question. In other words, observability$_2$ of all the concepts occurring in a potential answer is a necessary condition of its availability.

However, it can be required that all answers containing observable concepts be available, subject of course to possible other restrictions on the availability of answers. This variant will be called observability$_1$. It amounts to treating observability as a kind of sufficient condition of the availability of answers.

Both senses of observability are quite different from identifiability. As the notion of identifiability is defined here, it shares certain features with what is in the philosophy of science called measurability. Tentatively, and with certain important qualifications, identifiability could perhaps be considered a rational reconstruction of the idea of measurability.

In earlier discussions of observability and measurability, these notions are often assimilated in effect to one another. It is therefore important to heed the differences that obtain between them. For instance, observability (as I have defined it by reference to the interrogative model) is independent of any particular theory that the Inquirer may set up. In contrast, identifiability is a "theory-laden" notion. Not only is it defined so as to be relative to a theory; it is sensitive to the choice of the theory.

13. IDENTIFICATION AND UNCERTAINTY

These suggestions offer indications of the potential uses of the interrogative model and of the notion of identifiability that can be defined by its means. As a final whetter of the reader's appetite, I shall mention one more possibility for further development.[19]

Consider once again, a model M of a theory $T = T[P, Q]$, which is now assumed to contain two one-place predicates P and Q, whose identification is at issue. For some individuals $b \in do(M)$ one can establish

(1) $M: T[P, Q] \vdash Pb$

without using either P or Q in Nature's answers. For others, one can likewise establish

(2) $M: T[P, Q] \vdash \sim Pb$

For still others, neither (1) nor (2) can be established. The class of such individuals is called the uncertainty range of *P* or *UNC* (*M:P*). The uncertainty range of *Q* is defined likewise. The uncertainty range thus is, in effect, the range of entities in which a concept *P* cannot be identified. (More strictly, in the special case considered here, is something missing here? cannot be identified in terms of concepts other than *P* and Q.) As was suggested in the preceding section, we might also speak of the range of cases where a concept cannot be measured.

Now it can happen depending on what the theory *T*[*P*, *Q*] is like, that the two uncertainty ranges *UNC*(*M:P*) and *UNC*(*M:Q*) vary inversely when we move from one model *M* of the theory to another, say *M**. In other words, when one uncertainty range gets smaller, the other one gets larger. (How these relative sizes are measured need not concern us here.) The more accurately you can measure one concept, the more uncertainty there is with respect to the other.

This, of course, is precisely the phenomenon instantiated by Heisenbergian uncertainty in quantum theory, except that we are there dealing with the measurability of two time-dependent functions which assign to each particle a position and a momentum. Thus a theory of identification opens a new way of approaching of the problems connected with Heisenbergian uncertainty. For instance, we can see at once that the presence of such uncertainty can in principle be seen from the logico-mathematical form of a theory, just as nonidentifiability of a parameter in an econometric theory can be seen from the mathematical structure of the theory. (Cf. sec. 2 above.)

14. THE PLACE OF IDENTIFICATION IN INTERROGATIVE INQUIRY

From what has been said, it might appear that identification problems represent one particular application of the interrogative model of inquiry and of the techniques of inquiry that can be formulated by reference to this model. In reality, the theory of identifiability developed here is an integral and central part of the theory of interrogative inquiry itself.

In order to see in what sense this is the case, let us recall that the types of interrogative inquiry so far studied here are of a particular kind in that the inquirer's aim is to prove a given predetermined conclusion C. This is a very special and restricted typed of inquiry. It may occupy a privileged position in the pure theory of interrogative inquiry in that properties determine the logical properties of other kinds of inquiry. However, it is by far not the most interesting or important type of inquiry in practice. In the most interesting cases, the aim of an inquiry is to answer a question rather than to establish a conclusion already known. How can such types of inquiry be conceptualized?

The answer is fairly simple for propositional questions. For instance, to answer a yes-or-no question "*B* or not-*B*?" by means of an interrogative inquiry, the Inquirer can try to derive *B* interrogatively and independently try to derive $\sim B$ (in both cases from the same given theoretical premise *T*).

But what about wh-questions? For instance, suppose the question to be answered through interrogative inquiry is "Who are the *P*'s?", in other words, "Who (or what) are the individuals *x* (members of the domain $do(M)$ of individuals of the model *M* in which the inquiry takes place) which satisfy '*Px*'?" This question is obviously the same as: How can *P* be identified on the basis of *T* in *M*?

This question falls in two parts: Is *P* identifiable in *M* on the basis of *T*? If so, what is the identifying pseudo-definition of the form

$$(\forall x)\,(Px \leftrightarrow D[x, a_1, a_2, \ldots, a_k])$$

and what in particular is the *definiens* $D[x, a_1, a_2, \ldots, a_k]$?

This relates in a sense wh-questions and ways of trying to answer them to interrogative inquiry via the notion of identifiability. However, it does not yet enable the Inquirer to try to answer a wh-question, say "Who are all the *P*'s?" by means of a single process of interrogative inquiry.

It is here that the Extended Interpolation Theorem and its consequence, Extended Beth's Theorem, become crucial ingredients in a general theory of interrogative inquiry. We have just seen that wh-questions are in the theory of interrogative derivability to be construed as questions of identifiability. Now the Extended Beth's Theorem shows how such questions can be answered by a single process of interrogative inquiry. In order to find out whether *P* is identifiable in *M* on the basis of $T[P]$ the Inquirer can try to derive

$$(*) \qquad (\forall x)\,(Px \leftrightarrow P'x)$$

interrogatively from $(T[P]$ & $T[P'])$ in *M* without the help of answers containing either *P* or *P'*, where *P'* is a new one-place predicate. Moreover, if (*) is so derivable, from a closer examination of the proof of the Extended Interpolation Theorem it can be seen how the *definiens* can be read off from the interrogative argument that led to (*).

This, then, is how one can try to answer wh-questions by means of interrogative inquiry.[20] In order to answer a question like, "Who are the *P*'s?, the inquirer can try to derive (*) interrogatively from ($T[P]$ & $T[P']$) in the "world" M to which the inquiry pertains. If the attempt succeeds, from the resulting argument one can construct the *definiens* which yields the desired answer: the *P*'s are precisely those individuals which satisfy

$$D[x, a_1, a_2, \ldots, a_k]$$

This result shows convincingly the scope of the technique of interrogative inquiry.[21]

Department of Philosophy
Boston University

NOTES

[1] For a more complete treatment of the subject, see V. Rantala's *Aspects of definability*, Acta Philosophica Fennica, Vol. **29**, No. 2 (1977), North Holland, Amsterdam.

[2] See Cheng Hsiao, 'Identification', in Z. Griliches and M.D. Intriligator (eds.), *Handbook of Econometrics*, Vol. **1**, Ch. 4, North Holland, Amsterdam, 1983; F. Fisher, *The Identification Problem in Econometrics*, McGraw-Hill, New York, 1966.

[3] See H. A. Simon, 'The Axiomatization of Physical Theories', *Philosophy of Science*, Vol. **37**, No. 1 (1970), pp. 16—26.

[4] See H. A. Simon, *op. cit*; 'The Axioms of Newtonian Mechanics', *Philosophical Magazine*, Series 7, Vol. **33** (1947), pp. 888—905; 'The Axiomatization of Classical Mechanics', *Philosophy of Science*, Vol. **21** (1954), pp. 340—343; 'Definable Terms and Primitives in Axiom Systems', in L. Henkin, P. Suppes and A. Tarski (eds.), *The Axiomatic Method*, North-Holland, Amsterdam, 1959; J. C. C. McKinsey, A. C. Sugar and P. Suppes, 'Axiomatic Foundations of Classical Particle Mechanics', *Journal of Rational Mechanics and Analysis*, Vol. **2**, pp. 253—272; J. C. C. McKinsey and Patrick Suppes, 'Transformations of Systems of Classical Particle Mechanics', *ibid.*, pp. 273—289; P. Suppes, *Introduction to Logic*, Van Nostrand, Princeton, 1957, chapter 8; M. Jammer, *Concepts of Mass in Classical and Modern Physics*, Harvard U.P., Cambridge, 1961, especially ch. 9.

[5] E. Mach, *Die Mechanik in ihrer Entwicklung*, F. A. Brockhaus, Leipzig, 1883.

[6] See, e.g., J. Hintikka, 'The Logic of Science as Model-Oriented Logic', in P. Asquith and P. Kitcher (eds.), *PSA 1984*, Vol. **1**, Philosophy of Science Association, East Lansing, MI, 1984, pp. 177—185; 'Knowledge Representation and the Interrogative Approach to Inquiry', in M. Clay and K. Lehrer (eds.); *Knowledge and Skepticism*,

Westview Press, Boulder, CO., 1989; 'What Is the Logic of Experimental Inquiry?', *Synthese* vol. **74** (1988), pp. 173—188.

[7] See E. W. Beth, 'Semantic Entailment and Formal Derivability', *Mededelingen ven der Koninklijke Nederlandse Akademie van Wetenschappen*, Afdeling Letterkunde, N.R., Vol. **18**, No. 13 (1955), Amsterdam. This technique is formally speaking but a mirror image of a version of a Gentzen-type sequent calculus.

[8] There are of course other types of interrogative inquiry in which the aim of the game is not to prove a predetermined conclusion but to answer a question. I shall return to this matter in sec. 14 below.

[9] For these results and for their background see, e.g., J. Barwise (eds.), *Handbock of Mathematical Logic*, Part D, North Holland, Amsterdam, 1977.

[10] See my paper, 'Knowledge Representation and the Interrogative Model of Inquiry', in Marjorie Clay and Keith Lehrer (eds.), *Knowledge and Skepticism*, Westview Press, Boulder, Colorado, 1989, pp. 155—183.

[11] W. Craig, 'Three Uses of the Herbrand-Gentzen Theorem in Relating Model Theory to Proof Theory', *Journal of Symbolic Logic*, Vol. **22** (1957), pp. 269—285.

[12] E. W. Beth, 'On Padua's Method in the Theory of Definition', *Indagationes Mathematicae*, Vol. **15** (1953), pp. 330—339.

[13] L. Svenonius, 'A Theorem About Permutation in Models', *Theoria*, Vol. **25** (1959), pp. 173—178.

[14] E. W. Beth, *op. cit.*

[15] This theorem was first formulated in Jaakko Hintikka and Stephen Harris, 'On the Logic of Interrogative Inquiry', in A. Fine and J. Leplin (eds.), *PSA 1988*, Vol. **1**, Philosophy of Science Association, East Lansing, MI, 1988, pp. 233—240.

[16] See C. C. Chang, 'Some New Results in Definability', *Bulletin of the American Mathematic Society*, Vol. **70** (1964), pp. 808—813; M. Makkai, 'A Generalization of a Theorem of E. W. Beth', *Acta Math. Acad. Sci. Hungar.*, Vol. **15** (1964), pp. 227—235.

[17] For this issue, see, e.g., J. Giedymin, 'Logical Comparability and Conceptual Disparity between Newtonian and Relativistic Mechanics', *British Journal for the Philosophy of Science*, Vol. **24** (1973), pp. 270—276; David Pearce, *Roads to Commensurability*, D. Reidel, Dordrecht, 1987, pp. 154—158.

[18] See, e.g., R. L. Causey, 'Theory and Observation', in P. Asquith and H. Kyburg, Jr. (eds.), *Current Research in the Philosophy of Science*, Philosophy of Science Association, East Lansing, MI, 1979, pp. 187—206; R. Tuomela, *Theoretical Concepts*, Springer-Verlag, Wien & New York, 1973.

[19] See here J. Hintikka and P. Sibelius, 'Identification and Heisenbergian Uncertainty', forthcoming.

[20] Notice that in the principal wh-question to be answered through inquiry the wh-question is in effect given what I have called the universal reading, while the "small" questions through which it is answered is given an existential reading. (See here Jaakko Hintikka, *The Semantics of Questions and the Questions of Semantics, Acta Philosophica Fennica*, Vol. **28**, No. 4, Societas Philosophica Fennica, Helsinki, 1976, especially ch. 4.) This is as it ought to be. The Inquirer is trying to extract the maximal information from the questioning procedure, and hence prefers the universal reading of wh-questions. Nature is operating with the contrary purpose, and hence chooses the reading of wh-questions on which they have the least informative answers.

[21] In writing this paper, I have profited greatly from cooperation with Stephen Harris.